Rampage

Deckplate Leadership Learned under Sail

By

Brian Boland

Also by Brian Boland

CARIBBEAN'S KEEPER: A NOVEL OF VENDETTA
GRAVES IN THE SAND: A COLE WILLIAMS NOVEL
SIGNAL ON THE HILL: A COLE WILLIAMS NOVEL

Praise for Boland's books:

The novel is a fast read and a view of a life and experience not known to many Americans. Cole's story could be used to set up a series of books about him, certainly centered on the little known battles faced by the Coast Guard in protecting our southern border.

—Paul Lane, *Palm Beach*

Boland's extensive experience in the Caribbean allowed him to write Caribbean's Keeper with great detail. That's the beginning of a dark road that leads Cole Williams to a criminal world where failure to do right by his superiors could result in a punishment far worse than being fired.

—Chris Day, *Daily Advance*

In Caribbean's Keeper, Boland spins a story born from more than a decade of his own experience fighting the war on drugs.

—*The Norwich Record*

I recommend it for anyone who wants an adventure with a likable rogue (despite his faults) who is in it for the thrills, only to discover that underneath it all, he actually has a conscience. Great story.

—Ian Wood, *Novellum*

Rampage

Deckplate Leadership Learned under Sail

By

Brian Boland

WARRIORS PUBLISHING GROUP
LOCKHART, TEXAS

To Elli: I hope you'll always love the sea.
Love, Dad

Chapter 1

Tucked against low dunes and sitting on the damp trampoline of a catamaran, I stared out at the Chesapeake Bay. A shapeless black mass underneath the warm summer air, the dark in front of me was pierced only by the evenly spaced lights of the Chesapeake Bay Bridge-Tunnel that ran north, far beyond an invisible horizon. I could hear dwarfed waves roll in from the Atlantic Ocean to the east, stretched thin as they ran west around Cape Henry only to be tamed again as they wrapped further south across shallow sandbars before finally spilling over and against the narrow sand at my feet. Gently rolling and rhythmic whitewater, briefly visible under the moonlight, ran up the shore only to disappear, the saltwater drawn out once more into the darkness. To me, the ocean had always been a living and breathing thing. I sat on a stretch of sand called Chic's Beach, the place where I'd spent my youth, and was wholly consumed by the thought of leaving here in the morning.

Wrestling with feelings I wasn't yet equipped to understand, I knew only that I'd be parting ways with home. In the years that followed, I would grasp those complex emotions more firmly, but in that moment, I couldn't quite nail down whether it was excitement, nerves, remorse, or regret. In time I would realize that before any major journey in life, each of those emotions mixes with one another, swirling like rum, Coke, and a little bit of lime, and when taken together, the effects were intoxicating. Something new, something scary, something unknown. But perhaps most importantly, it was something that I wanted. The same complex part of the human psyche that had drawn sailors to the sea and astronauts to the moon was drawing me towards something as well. Everyone feels it in their life, but only a select few see those dreams come to fruition. It was that same burgeoning restlessness that has led many an explorer to great fame and some to untimely deaths. Neither success nor failure were assured. I couldn't be sure if I liked the feeling or dreaded it, but it was a feeling that was not going away. In the morning, I'd cross the Bay Bridge-Tunnel enroute to the United States Coast Guard Academy, where in just over 48 hours, I'd be reporting in as a member of the class of 2003. In the grand scheme of things, it was a small step to take, but nonetheless one that would shape the rest of my life.

On the beach beside me sat a girl who was herself wholly consumed with attempts to draw from me some assurance that we'd stay in touch. To her dismay, I made no such promise, reckoning in my restless mind that the next four years would leave little room for hometown romances, especially those born haphazardly on the eve of a young man's departure for military service. Clearly disappointed, she drove me back to my house and the night ended, well after the curfew imposed on me by my parents. Having been waitlisted by the academy admissions department for some time, it had been just over a month since I'd been tendered a full appointment, and I'd correctly wagered that, given the particulars of my situation, no punishment awaited me if I stayed out too late.

Tracing its roots back to 1876, the Coast Guard Academy sits on roughly 100 acres of land on the west bank of the Thames River, at the southeast corner of Connecticut. The dark water of the Thames is drawn from a watershed that encompasses both Connecticut and parts of Massachusetts, fed by both the Yantic and Shetucket River before they combine to run south, past the city of New London, then into Long Island Sound. It runs with enough consistency to avoid icing over for all but the coldest days of a New England winter. Further south, and across the Sound sits the eastern tip of Long Island, with the North Atlantic Ocean beyond.

The academy itself is a scenic campus with old trees, a large and manicured parade field, and mostly red brick buildings in the classic New England style. To anyone but an 18-year-old mere minutes from reporting in with a freshman class, it is a picturesque campus. To the unfortunate few about to embark on their four-year journey towards a degree and a commission, it is nothing short of wildly intimidating. Chase Hall is the most foreboding building on the campus, serving as the barracks, or dorms, where all the students live during their four years as cadets. The peculiar rules by which cadets, and more specifically the freshmen known as Swabs, abide by while inside Chase Hall might seem cruel and unusual to some, but a spartan existence with an emphasis on physical discipline is a time-honored tradition at military academies around the world and has a proven track record of quickly transforming young men and women into capable military officers.

I brought little with me on the 6th of July 1999, all of my possessions fitting into a small suitcase with plenty of room to spare. Sitting with my parents on a grassy embankment next to a parking lot, we waited for my designated time to report in. The wait was unpleasant as it served no purpose other than to delay the inevitable. I didn't want to say goodbye, and yet at the same time, I would soon need to. I knew beyond a doubt that the next few days and weeks would be miserable and trying, but the unending wait for it to begin now seemed even worse than just getting the damn thing started. At the time, I couldn't put myself in my parents' shoes, but looking back now, I have little doubt that they were feeling many of the same things.

When my time came, I carried my suitcase with the few items I'd been told to bring, namely white socks, white underwear, and white V-neck t-shirts, and checked in with some senior cadets at a folding picnic table before stepping rather unceremoniously into the 'quad' which was a concrete open-air space surrounded on each side by towering brick walls. With me, and as instructed, I also had an alarm clock, although the AM/FM radio feature would be off-limits until a future yet-to-be determined date. Beyond those essentials, I'd brought two framed pictures, one of my father and me, the other my mother with the family dog. In a small photo album, I'd gathered a random selection of photographs as well, with the foolish hopes of delaying the inevitable loss of memories of my up-until-now civilian life.

With parents still on the campus, all of whom were craning their necks to see what torment awaited their precious little darlings, the first day was an uneventful affair. If we were yelled at, it was with muted voices and only briefly so as not to upset the spectators and, more importantly, future donors to the Parents' and Alumni Associations. Within the first few hours, I wore a badly wrinkled uniform and was marching horribly out of step, paraded around the grassy expanse with a band playing somewhere from the sidelines. We were then given some time to say goodbye to our parents before being rounded up and quickly marched away.

Until we were given a full medical evaluation, we weren't allowed to run, so the first week or so wasn't all that bad. There was yelling and marching, followed by more yelling and more marching after that. Then we marched some more and got yelled at yet again to march better. Three meals were provided each day, which quickly became a test to see if we could get out of the calorie deficits we were all surely suffering from. We were each issued a 'Running Light' which was a pocket-sized book full of incredibly small font that educated us with information about the Coast Guard, the Coast Guard Academy, and how we were to conduct ourselves. Everything from *Coast Guard Vessels and Aircraft* (pages 57-81) to *Phone Etiquette and Writing Messages* (pages 127-128) were covered. Minutes after being handed our 'Running Lights,' we were told that we needed to immediately memorize all 172 pages. A moment or two later, our cadre, the second-class (junior year) cadets were ripping into us for why we had not yet memorized the contents of the little books we'd just been handed. At times it was intimidating, and at others the entire ordeal bordered on the comical.

My first letter home, dated July 14th 1999, started out, *The other night we had a ceremony where we married our rifles.* I then explained to my parents how I'd named my rifle *Pamela Lee,* after the star of the non-critically acclaimed television show, *Baywatch*. It was a somber ceremony late one night where each of us Swabs were given a rifle and then marched up a dimly lit hallway to a table where several cadre were seated. We then stood at attention and were asked for the name we'd chosen. Hopefully I drew some laughs, but I can't recall. Afterwards, each of us were then sent back to our rooms with our rifles and told to sleep with them in our beds for the night. Truthfully, Pamela Anderson bore much of the responsibility for my choosing to enter the Coast Guard, as I had watched the show throughout high school and had been intrigued with the cameo appearances of Coast Guard boats and helicopters. I would find out in the coming years that Hollywood often does a tremendous disservice in providing an accurate depiction of military life.

I closed the letter asking for packaged food and tried to assuage the fears my parents were no doubt harboring by telling them I'd gone to church. In reality, Sundays afforded us the opportunity to attend the services of our chosen faith. Many of us quickly renewed our relationship with god, at least for those first few weeks. In addition to not being yelled at for an hour or so, I discovered that the clergy, from both the Catholic and Protestant faiths, set out a spread of orange juice and snacks to nourish their flock after their respective morning services. Within a week, I was attending both the Catholic and Protestant services, one immediately after the other, reaping the rewards of unlimited donuts and blueberry muffins. Church itself was a miserable experience, as most cadets sat around trying to hide the tears that were welling in the corners of their eyes, but the donuts were good and provided some free calories to prep for the Monday morning workouts looming over all of us.

For an incoming class, that first summer is referred to as Swab Summer, harkening back to the nautical term for a mop. There I was, no longer an individual, but now part of something greater than myself. I had become a mop and spent a great deal of time reflecting on all the other mops that had come through those gates over the past 100-plus years and how they must have felt. One key component to military indoctrination is the process of breaking down individuals so that they no longer thinks in terms solely of themselves. Military history is full of examples where individuals did great and heroic things not for themselves, but for the greater good. It could be argued that this is not a normal human instinct, so the military finds creative ways to build young recruits up into the desired mold. It did not take long for them to break me down. I do, however, believe that the building-back-up process is the far more complex of the two, and as I would find out, that latter process played out much slower than the former and in entirely different ways.

Each day, we were up early in the morning, doing calisthenics and running in formation through the chilly New England air. After a quick breakfast, our days were a mix of classroom lectures, obstacle courses, and seemingly unending hours of pushups and circuitous formation runs around the academy grounds. Throughout each day, our eyes were to be kept "in the boat" meaning we were not supposed to look around. At anything. Whatever was directly in front of us, that was what we were supposed to look at. Thankfully, the long runs provided brief windows of time to look around the campus and take in the sights.

Perhaps more than any other single activity including sleep, we marched. Marching, once you get the hang of it, isn't all that bad. But learning can be a challenge to the uninitiated, made even worse by our cadre. For the Swabs, they were akin to drill instructors. In reality, they were a mere two years older than us, but most were well-versed in screaming, the art of making angry faces, the appearance of an always meticulous uniform, and other general acts of intimidation. Day after day, we marched with our rifles and learned the manual of arms as our cadre barked out commands in front of us.

We carried M1 Garand rifles. With each command from our cadre, we were to perform some ceremonial twist or pivot of our unloaded rifles, all of which culminated with a manual cycling of the action. We would push the charging handle back, lock the action and await the next command. Soldiers at war carried *en bloc* clips loaded with eight rounds of 30-06 Springfield ammunition, which were inserted into the open action and the bolt would then close and load a round. We had no clips with us, so we were instructed to mash our thumbs down to the slide and follower to release the bolt. I knew little about guns at the time or that the M1 Garand had been a favorite of General Patton during the Second World War.

Designed by John Garand, the rifle had been a critical weapon for the American GIs as they fought all over the world. As I found out, it also had a unique propensity to violently pinch one's thumb when the operating rod slammed forward and the bolt shut. Lost on us at the time also was the fact that firearms technology had come a long way since 1945. Modern rifles were lightweight and short in overall length to facilitate ease of use. The M1 Garand was long and heavy, made even heavier by the lead that had been poured down all of our barrels in order to render the weapons inoperable, lest one of us get any bad ideas.

The long days of physical conditioning in early July didn't bother me. After receiving my official appointment, I'd spent the previous month running miles on Chic's Beach each morning to prepare myself. My father had attended the U.S. Naval Academy and graduated with the class of 1973, so I had some insider information, albeit slightly dated, that the ability to run for long periods of time would be of tremendous benefit. No single piece of advice could have prepared me better for that first summer.

As the days wore on, it became apparent that at some future date, we'd have the opportunity to spend some time at the academy waterfront. For the first few weeks, I'd catch glimpses of it during our runs or long marches around the campus. Down on the Thames River, an armada of sailboats awaited us, small FJs and 420s that were pulled up neatly on the floating docks. Beyond that, a fleet of J-22s floated quietly in their slips, several Luders Yawls were tied to a dock just south of them, and lastly a random selection of larger boats just further to the south on another dock caught my attention. Having grown up on and in the water, I missed it terribly during those first few weeks. I knew nothing about sailing, but as I ran in formation during those humid July days, I was quietly dreaming of the opportunity to somehow get on those boats.

At last, towards the end of July, we were marched down to the waterfront in our goofy boat shoes and dropped off for the afternoon with a new cadre. A long wooden causeway led out to the sailing center, a white building perched on Jacob's Rock. These 'waterfront cadre' were the same class of juniors as the ones that had run us into the ground up the hill, but they seemed far less menacing in their khaki shorts and t-shirts as they tried to lecture us on the basics of seamanship. Sailboat nomenclature and the tying of basic knots took up much of the first lecture. In short time, we were paired up, outfitted with life jackets, and given sails. Down and onto the floating docks, these cadre then helped us rig up our dinghies before we pushed off into the Thames.

In the heat of the summer, any real breeze was nonexistent as we floated around with our sails mostly luffing in the idle wind. But a point came when my sails found some wind, the boat heeled over just slightly and looking aft, I saw a faint wake behind us. I was sailing. Adding to my great pleasure was the fact that no cadre were in sight to yell at us. For the most part, we were left alone for an hour or two on the river. Several of the cadre motored around in small skiffs, making sure none of us floundered on the rocks of the east bank of the river or drifted too far to the south with the current.

Many of my classmates had been recruited for specific sports. Several were accomplished high-school sailors and were therefore highly sought after by the sailing team. I quickly figured out which ones knew what they were doing and studied their movements from my floundering dinghy. At first I could do little more than mimic whatever actions they were performing, but it didn't take all that long to sort out the basics. Keeping the bow just off the wind kept the sails full and hiking one's own body weight out to windward to counter the heeling of the boat yielded even better speed. Holding the tiller, I controlled the rudder and worked slowly to smooth out and steady my course up and down the river. Those with considerable experience made the act of sailing look smooth and controlled, each of their movements a choreographed routine whereby they wasted the least amount of the boat's energy and surged ahead of the rest of us. I was no doubt flailing around and unnecessarily spilling wind from my sails. But I was sailing.

I learned that by pulling in the main and jib sheets in stronger winds, the sails could be loaded up and their energy transferred to moving the boat even faster through the water. A centerboard was pushed down through a narrow slit in the middle of the boat, adding some countering moment against the constant tendency to want to roll over on its side. Tacking was turning through the wind and a quick shift from one side of the boat to the other kept the small dinghy working upwind. Downwind was similar, although the jibe maneuver knocked many of my classmates into the water as the boom swung wildly across the cockpit. Because of this, the small clinic on base served as the Coast Guard's pre-eminent facility for quickly stitching up head lacerations. At the time, checking for signs of a concussion were not as in vogue as they are now, but I have little doubt that there were a fair number of rattled brains that summer. Years later, the academy began to require swabs to wear helmets while on the water and this was no doubt a welcome step to preserve the mental capacities of future Coast Guard leaders.

That day, I managed to stay in my boat. And with each subsequent day on the water, I learned as much as I could, trying on each occasion to go faster and farther with less clumsiness through each tack and jibe. On at least one balmy afternoon, I jumped in the river and held onto the rudder with both my hands, my classmate still in the boat, holding the sheets for both the main and jib while looking at me with confusion on their face. I didn't care. I felt the boat pull me along, the cool water running around and over me as I dipped my head fully into the Thames and let the water rush over my neck and down my back. I was hooked.

For the rest of the summer, we were on the water every few days and with a little experience under our belts, the cadre put together a regatta for us. In my first race, I wrote to my parents in late July that I'd finished fifth. I considered my ability to successfully navigate a racecourse without grievous bodily injury to be a significant accomplishment. At the end of each session on the water, we pulled the boats back up on the docks, rolled and neatly put away our sails, then formed back up for the long march up the hill to Chase Hall, our more sinister cadre waiting patiently to remind us that this was no summer camp. I dreaded those marches, often sneaking a look back at Jacob's Rock and torturing myself with daydreams of just one more hour on the water. Those days were the first glimmer of hope that I felt. Perhaps I'd survive the next four years. And if I did, I was all but certain that the waterfront would be my sanctuary.

Chapter 2

'Swab Summer' was just over six weeks long and hadn't been all that bad. It was a busy routine that allowed for almost zero free time, and I was fine with that. Physically, I felt I'd done well. Where others dreaded the long runs and endless pushups in the hallways, or 'passageways' as we were instructed to call them, I found that I enjoyed the conditioning. But as the summer wore on, we'd taken several placement exams to flesh out the floaters and sinkers for the academic year. I could run and do pushups with the best of them, but academically I was not confident I'd be able to hold my own. I'd done fine in high school, but I was far from an overachiever. In my senior year, I'd dropped out of trigonometry and opted for an elective where I helped put the yearbook together with a gaggle of pretty girls from my graduating class. Tactically speaking, this had been a good move, but strategically it was a costly blunder. In terms of setting oneself up for success at the Coast Guard Academy, trigonometry would have been the more prudent choice. I rather quickly surmised that my skills in cropping photos were not going to do much for me at the collegiate level.

Most of my classmates had taken pre-calculus, some of them having attended civilian colleges for a year or two in an attempt to take as many math and science classes as possible prior to reporting in. During the calculus placement exam, while most were furiously working equations with their pencils, I scratched my head trying to figure out why there were letters in math problems. All around me my peers were studiously working their way through the test while I was trying to figure out what exactly the letter 'x' was supposed to mean numerically.

In August, our school year began and by the 12th, I cheerfully reported to my parents: *We had a calculus quiz yesterday which I failed. Fortunately, I wasn't the only one nor did I have the lowest grade so right now I feel a bit above scum.* From that point forward, I developed a habit early on which served me quite well for the next four years. I routinely sought out teachers during their office hours for help. And to their credit, and the academy's as well, many put in countless hours trying in vain to help me. Their efforts were rarely successful. However, it is likely that my persistence may have yielded some advantages on the end-of-semester grading curves, if for no other reason than most of those teachers preferred to not have me relentlessly knocking on their door for another semester.

With the school year also came the fall sports season. I'd signed up to sail and at the designated time made sure I was down at the waterfront for whatever program I could get involved with. There was a varsity dinghy team, the Same dinghies we'd meandered around the Thames on during Swab Summer, and a varsity offshore team, crewed mostly by more senior cadets. I was unceremoniously lumped in with the junior varsity team but happy nonetheless for the opportunity to be on a boat.

On the first of September, I wrote to my parents: *I got to sail on a ludar (sic) today. I was on Blue Goose. Do you remember pointing that name out when we were on the docks? It was fun. We have three races and two of them are overnights. I can't wait.*

What I meant to say was that I had sailed a Luders Yawl. These were old 44-foot boats with what appeared to me to be classy lines. Amidships at their widest point, the rails seemed to slope gently down from a slightly raised bow and stern. There was a small open cockpit aft and a cabin with a kitchen, navigation station, sail storage, and racks for overnight sleeping arrangements. With inexperienced cadets at the controls each year, the four Luders yawls were in no way kept in boat show-quality condition, but they were practical and equipped with all the essentials one would need for a weekend of open-water sailing.

As inexperienced as I was, they presented an image of sea-going stability with their heavy-duty oversized rigging and a mildly worn, yet cared for, look. What I didn't know at the time was their history. Tracing back to their introduction at the U.S. Naval Academy just after the Second World War, they'd been designed by Bill Luders, originally as an all-wooden hull with the fiberglass versions coming some 25 years after the introduction of the first wooden hulls. The Naval Academy had parted ways with them entirely, but four had been built specifically for the Coast Guard Academy and had been maintained over the decades by the waterfront staff, a mix of civilians and enlisted men and women.

They were indeed stable boats, not unlike a floating Sherman tank with sails. They were so stable that the Coast Guard Academy had no qualms about sending a group of young inexperienced fourth-class cadets out onto Long Island Sound for an afternoon of sailing; there were few if any things that a Luders yawl couldn't hit and carry on as if nothing had happened. Out on the Sound, we were akin to a bulldozer under sail. At least one of my classmates put that to the test that season, sailing back to the academy where he reported a head-on collision with another boat. A thorough inspection revealed no damage, and the boat was quickly cleared to sail again.

When underway either for practice or a weekend regatta, we had a safety officer with us, generally a Lieutenant or Lieutenant Commander, who seemed content to offer advice and some basic coaching from a comfortable seated position somewhere near the stern. For the officers, it was a break from their duties as faculty up the hill and was no doubt one of the perks of their job. We did the grunt work, and the boats were forgiving enough in the fall breeze to hold their course while we learned the ins and outs of practical seamanship. Too heavy and old to competitively race, the four Luders yawls were mostly left out of the varsity circuit and held in reserve for us youngsters to sail.

Nearly 11 feet at their widest point, the boats had a main mast just over 50 feet in height and weighed nearly 25,000 pounds. Aft of the rudder post was a smaller mizzen mast (this defining feature made it a yawl), which many sailors will argue to this day provides far more in terms of overall appearance than sailing efficiency. Drawing six feet, the boats had a heavy keel containing nearly 10,000 pounds of ballast. That long keel led aft to a single propeller enshrouded by a large rudder. Her dimensions came together to offer just over 1,000 feet of sail area.

My lasting impression after feeling the boat under sail was how easily she could absorb an unexpected gust of wind. Sailing dinghies had taught me how susceptible a boat was to capsize, but out on the Sound, the yawl's main and jib could load up with a stiff breeze and heel gently to one side, the 10,000 pounds of ballast in her massive keel easily countering the wind to keep the boat under control. As I'd learned on the river, smaller and lighter boats, with their diminutive centerboards, tended to roll up into the gusts when they were overloaded, pointing nearly straight into the wind, and quickly losing their speed as the sails luffed and snapped like a whip in the stiff breeze. It was constant work for a dinghy crew to keep the boat under control. At their fastest, those dinghies hung precariously on the edge of capsizing, their coxswain and crew extending their bodies out over the windward rail to keep the boat upright. With each small change in the wind, whether it be direction or velocity, the dinghies required immediate action, or they'd quickly fall off course and lose their speed.

The Luders were entirely different boats. Through the gusts that worked their way down the Sound, the Luders could hold a steady and constant course. From this, I learned to look ahead and upwind for each gust, often first appearing as a darkened stain on the water moving towards us before the new cool breeze would hit my face and fill the sails. Each time the rigging would creak and groan, the already-overused sails stretched once again to their limit, but the Luders seemed undaunted. As the sails filled further, the boats would load up and heel slowly to the leeward side. Moments would pass before the boats would then accelerate as the energy captured by the sails transferred to the heavy hull moving steadily through the water. It was as if they could somehow absorb and store that raw energy that the dinghies seemed to struggle with so much.

As a gust would pass, the Luders would roll gently back down, but their momentum carried them onward through the Sound, and if the gusts were timed right, the boats would hold a near-constant speed, their rigging and sails harnessing what they could as us young cadets held on, each of us learning to act in synchronous fashion and more importantly as a team. I was far too captivated by it all to realize the lessons right in front of us on those formative days of open-water sailing. Those first days laid the foundation upon which I'd build some semblance of nautical acumen, which would soon lead to bigger and better things on the sailing team.

Any helmsman had little control over the boat without proper action on the part of his teammates trimming the mainsail and jib. As an individual, he was powerless through a tack or jibe without help from the half dozen classmates scurrying around the boat, loading the leeward jib sheet, trimming the main, and helping the jib around the mainmast. Leadership at that point mattered little, as this was purely an exercise in teamwork. What made the lesson stick was that it was not some lecture from a chalky blackboard, but rather we were active participants in the tangible results of half a dozen people working together towards a collective goal. Our hands hurt, no doubt some of us had earned some bruises, and we'd worked our way out of more than a few tense moments on the water. At the end of a day, slightly sunburned, dehydrated, and tired, I walked back up the hill to Chase Hall incredibly happy. I was also naively certain that, with the handful of hours I'd spent learning the ropes, I was now entirely equipped with the skills and nautical proficiency to sail one of the Luders around the world. Solo.

By 13 September, I reported back to my parents that I'd sailed a J/22 on a Monday afternoon. At 22 feet in length, and weighing around 1,800 pounds, the academy had a fleet of them for use on the river. On dinghies, I'd learned the basics of maneuvering a sailboat, tacking and jibing, and coordinating sail trim with the rudder. The Luders had been my introduction to sailing a big boat, which I'd quickly taken more of a liking to from the beginning. As an intermediate step between the two, the J/22s were a perfect platform to focus on the mechanics of the various evolutions needed to make a larger boat sail fast.

With a conventional rig and wooden tiller at the stern, the J/22s had a mast height of 25 feet and a sail area of roughly 225 square feet, which was proportionately much smaller, and more forgiving, than the Luders rigging. Most important for me, the J/22 rigging was simplified and the deck far less cluttered by cleats, winches, dorade boxes, and lines. The boats were also equipped with a fin keel that made them handle more like a dinghy than a larger sloop or yawl. The ease of maneuvering gave me an opportunity to study the intricacies of a coordinated tack or jibe. Out on the Sound, I had only been able to pick up bits and pieces of what we were doing as a team, but with practice on the J/22s, a clearer picture began to emerge.

The helmsman would often call the shots, counting down to a tack or jibe before initiating the maneuver with a slight push or pull of the tiller. To tack, he or she would bring the bow up into the breeze steadily until the boat sailed through the wind to take it on the opposite side. At the same time, once the jib had lost the breeze and began to luff, it was the job of the crew trimming to release the sheet from the now windward side and begin to pull the jib around the mast and trim it on the new side. For the mainsail, the trick was to have the sail properly trimmed going into the tack. As the boat lost speed through the turn, the mainsheet was eased towards the new leeward side to help the sail load back up and build speed on a new course.

The J/22s normally sailed with a crew of three and their movements were also coordinated to make best use of their weight to keep the boat steady. During a tack, from an upwind leg, one crewmember was an extra and would move quickly up and around the mast to take a position on the new high side and help add weight to the windward rail. In addition to trimming the jib, the second crewmember also had to move from one rail to the other. Lastly, the helmsman needed to move his or her weight at about the same time. If done correctly, all three crew would shift their weight outboard together which aided in loading the sails at the boat steadied on a new course. If any of those individual movements were not synchronized, the tack would be sloppy at best. At worst, the boat would slow, sometimes to a complete stop, and the helmsman would have to work furiously to get the bow back down and re-establish wind across the sails.

On a downwind leg, the J/22s provided my first introduction to the spinnaker. A large unwieldy thing, the 'spinny' or 'chute' could carry the boat at tremendous speeds in nearly any direction downwind so long as the wind was kept somewhere abaft the beam. It required a significant amount of effort on the part of the third crewmember to keep in trim. Moreover, it necessitated smooth input and coordination from both the helmsman and crewmember trimming the sail to keep the boat on course. Clear and concise communication between the two was critical. I would learn years later in aviation about the concept of a 'shared mental model,' which is a fancy way of ensuring that all members of a crew are on the same page with both what's going on in the moment and what's going to happen in the near future. Sailing is no different than two people flying an airplane, the consequences however can be a bit more serious in the sky. On the J/22, if we weren't synced up or if two people were chasing separate outcomes, the sails would luff, our speed would decrease, and we'd end up floating to a complete stop. Words would be exchanged, and we'd get moving again. In aviation, if the airflow stops over a wing, this leads to a stall in a three-dimensional environment with no option to float for a minute or two to get one's bearings.

To rig the spinnaker, it was hoisted up with a halyard to a position high up on the mast, then a long pole is attached to the mast at about head-height to hold out the windward clew via a line called the guy. The leeward clew is then controlled by another line, or sheet, and operates in a similar fashion to a jib sheet. When it works, a downwind leg can be an exhilarating experience. When it goes bad, as was often the case for an inexperienced sailor like me, it can be a rather uncomfortable event. The natural tendency for a crew to cast blame on each other is a detriment to their efficiency as a team, but also an understandable reaction to a stressful situation. As a fourth-class cadet, I kept my mouth shut during these semi-frequent crew breakdowns and watched the blame be tossed the boat around like a hot potato. Eventually, cooler heads always prevailed, and we carried on with the practice. I learned a great deal about managing interpersonal relationships from watching those arguments play out. Yelling was rarely an effective method to resync a frustrated crew. This stood in stark contrast to the incessant yelling within Chase Hall and I developed a strong disdain for it early on in my cadet existence.

At some point in October, I'd spent a weekend away from the academy on a sailing trip. By a strange twist of fate, myself and a friend, Chuck, had been bumped up to the varsity offshore team for several regattas. Neither of us were sure what exactly had transpired, but we heard bits and pieces of a rumor that some third-class cadets on the team (a year ahead of us) had caused enough trouble to be permanently booted. There were whispers of some kind of excessive celebratory evening on Long Island that had gone awry and made its way back to the academy, necessitating some changes to the team. None of this mattered to me, but when approached by the coaches with an offer to hop on the varsity boats, I didn't hesitate.

Looking back on that first season, I attribute my elevation to the varsity team to a combination of attitude and effort. It was most certainly not a result of my raw talent. For that first season, I was often one of the first to practice and the last to leave the docks. If there was work left over or sails to be put away, I offered to help, often not knowing exactly what to do, but trying my best to figure it out. When a hurricane had threatened to make a run up the coast of New England, I'd hung around to help secure all the boats, not with the intent of impressing anyone, but more so because I was genuinely eager to learn all that I could about practical seamanship. Someone had apparently noticed. Chuck, on the other hand, had a sailing background. He also had an infectious smile and attitude like mine. His rise to the varsity team made sense, whereas I was not so sure I was ready. Nevertheless, we considered ourselves incredibly fortunate.

Practices with the varsity team were more intense. The upper-class cadets worked quickly and seamlessly, repeating evolutions over and over again, until the movements were fine-tuned, and the boats moved with ease around the triangular courses set up on the river. My first task was assisting with trimming the jib through tacks. It required keeping the sail loaded up under strain until the bow swung through the wind, at which point I'd release the jib sheet from the now-upwind side then scurry down to the leeward side and crank the winch until the sail was once again trimmed. From there, I'd leap back up to the windward rail and hike as much of my body up and over to help counter the heeling moment. Mine was a simple job and once I'd winched the jib in, it became the job of a more senior cadet to work with the helmsman and main trim to keep the boat moving.

Once comfortable with my position on the rail, I turned my attention to the jib, trying to glean as much as I could about proper shape and trim. When the wind would ease, so too would the sails, and in the gusts, they'd be trimmed in. A forestay ran from the bow to the top of the mast and a backstay ran from the stern to the mast, helping to keep the rig tight. What fascinated me most was tuning the backstay to optimize the shape of both the main and the jib. It seemed that each individual part of the boat worked in concert with all the others and after only a few weeks of practice, I had a good understanding of just how much there was to learn. The rig, while mostly aluminum and steel cables, was constructed in such a way as to be flexed and bent to make the most of the sails. It was all overwhelming for someone new to the sport.

Varsity regattas were two-day affairs, almost always with a Saturday night anchored or docked near the hosting yacht club. The upper-class cadets laid down some ground rules, namely that me and Chuck were not to leave the boat. Given the misery of Chase Hall, we each thought a night on a boat under the stars sounded pretty good, even without the shoreside shenanigans that we could clearly see developing at the yacht club. To sweeten the deal, one of the first-class cadets, Sean Krueger, brought a six pack of beer for us to share, with further instructions not to do anything stupid or tell anyone of our good fortunes. We were more than happy to oblige.

Sean Krueger would graduate and go on to fly H-60 Jayhawks. He was killed in a crash on July 7th, 2010, while flying in the Pacific Northwest in transit to Sitka, Alaska. The aircraft had impacted some improperly marked power lines off LaPush, Washington. Two other crewmembers perished while the co-pilot (another friend of mine from the sailing team) survived. In the view of many junior officers at the time, including myself, the copilot—and sole survivor of that crash—was then unfairly targeted by the Coast Guard as a scapegoat. The charges against him, including negligent homicide, dereliction of duty, and destruction of military property were ultimately dismissed, but the entire ordeal cast a dark shadow over the aviation community.

By the middle of October, I'd secured a spot on a trip down to Annapolis to race at the Naval Academy. While an unremarkable regatta, it was yet another weekend away from the academy and provided me the opportunity to hang out in civilian attire at night, as we were prohibited from wearing anything other than issued uniforms while on academy grounds or in the nearby area. Losing the 'privilege' of wearing jeans may not sound like much, but when you're 19 and forced to wear a uniform day in and day out, one quickly becomes nostalgic for the feel of walking around in an old pair of jeans and a t-shirt. Making the deal even sweeter, my parents drove up to meet us and took the entire team out for dinner.

The season closed out shortly thereafter as the New England winter descended on the campus. I buckled down and stumbled my way through Chemistry, Calculus, and a whole litany of unenjoyable classes, most of which I had little to no interest in. By late November, snow covered the grounds and the boats had been hauled up out of the water, wrapped and winterized until the spring thaw. The river rarely iced over, and I spent much time during those cold months staring out windows from up the hill down at the dark water running south into the Sound.

In the spring, one class interested me more than the others: Nautical Science. It was, in general terms, a class on seamanship and navigation. We were given instruction on plotting coordinates, the proper use of charts, and general ship-handling. There were elements of my first season of sailing that tied directly into the class and that seemed to resonate in my otherwise obstinate mind. Looking over the charts, I would run my fingers across the shoals and rocky points of New England, daydreaming and thinking back fondly on those early fall days where we'd sail out the mouth of the Thames and work our way further out towards Long Island and then east towards Rhode Island.

In my first year as a cadet, I had developed a very basic appreciation for the sea, more so than I had while surfing and swimming as a teenager. While we hadn't ventured too far from the coast, it was the furthest I'd ever been out in open water, and I enjoyed the challenges that came from those experiences. I'd developed an eye for the subtlest changes in the wind's direction or velocity, learned to dead reckon by reading the currents that swirled in and around Long Island Sound, and felt the pressure of coordinating the boat's movement with those of her crew. In short, I was developing those instincts that could not be replicated in a classroom lecture. Perhaps most important, I'd learned that sailing a boat only worked well when everyone was committed to acting as one cohesive team with a clear understanding of whatever task lay in front of us. I'd seen tacks and jibes go horribly wrong and watched that cohesive team quickly fall to pieces when things didn't go as planned. Too green to fully grasp the important lessons I'd learned, I was nevertheless entirely hooked on sailing.

Chapter 3

I took a train home for the winter break. It was far slower than flying, but with the unpredictability of winter snowstorms, my dad had recommended it as the safest bet for uninterrupted travel. A career as a naval aviator had instilled in him the importance of always planning for the worst case. I most certainly rolled my eyes at the suggestion but would come to greatly appreciate his logic during my own flying career. The train would take me back to New London at the end of our winter break regardless of the weather. Being late for anything in the Coast Guard, whether it be a rescue at sea or a Chemistry class, was frowned upon. The train was indeed slow, stopping at times along desolate stretches with little more to look at besides endless snow and woods. Making matters worse, each of those frequent stops lasted for what seemed like an eternity. But it got me home the following morning, where I had a two-week break for the holidays.

Waking up anywhere besides Chase Hall felt strange and while it was good, the return trip loomed over me and put a damper on my mood for the duration of the holidays. After New Years' Eve, I took the train back to New London. It was, as expected, far more brutal in that as each hour passed, I travelled further back into the cold and dark New England winter. Arriving around midnight, I took a cab from there to the academy, where I wandered down some dark and seemingly abandoned passageways until I found my room. Climbing into my sleeping bag, I fell asleep feeling remarkably bad for myself. I missed home all over again, the pit in my stomach all too familiar like the first day I'd reported aboard.

Learning the ropes onboard Eagle

Fortunately, there was little downtime to feel sorry for myself, as the spring semester began the next day. Within a week or two, we spent some time on *Eagle*, the academy's training ship that was moored down at the waterfront. Taken as a prize from the German Navy after the Second World War, we were told that it would teach us leadership. Having been built in 1936, she had an overall length of 295 feet and drew almost 18 feet of water. Compared to the Luders and J/22s, her 22,000 square feet of sail area was something to behold. In the frigid winter air, we practiced climbing up and down the rigging of her three masts until our fingers were numb. We were told this was done to prepare us for the summer when we'd sail her for several weeks at sea. I had learned by this point to take any such claims with a healthy amount of skepticism.

As the spring semester got into motion, our class had grown smaller. One by one, dozens of my classmates had been removed for either academic or disciplinary issues. Beyond that, there were many who had come to realize that the academy experience, or perhaps the Coast Guard, was not something they wanted to be a part of anymore. For some of them, perhaps it was the military in general, but my early sense was that many of the ones who left on their own found the thought of four years in such an institution unbearable.

I had wrestled with those feelings myself, having made the mistake during the fall of calling two of my high school friends on the Friday before Halloween. From a pay phone in Chase Hall, I heard music and girls in the background, and the two of them explained in great detail how they'd spent considerable time fashioning a cooler from an oversized pumpkin. The pumpkin was, at that point in our conversation, full of beer and ice. I, on the other hand, had taken a break from cleaning the common spaces, to include hallways, storage closets, showers, and bathrooms to call them and say hello. We talked for a few minutes and after hanging up, I realized that it would do me no good to talk to them anymore. It was a near-instantaneous decision on my part to sever whatever ties were left with my friends from back home.

Because I was struggling academically (a fact that shocked no one including myself), I constantly felt like an underachiever and that in turn meant that I had something to prove. I was determined not to quit. I'd somehow been accepted and told myself that it would take an entire platoon of officers to drag me off that campus. As soon as I stepped foot on the academy grounds, I vowed to never leave voluntarily no matter how deep or dark of a pit I found myself in. I felt as if my acceptance there was some clerical mistake, that I wasn't good enough to be a cadet, or an officer for that matter. Thankfully, I was stubborn as hell and that feeling only added fuel to the fire. It was entirely counterproductive, but I became angry with the people, and there were many, that told me I wasn't cut out for academy life. I was in no way trying to get ahead of anyone or stand out in the crowd, my focus was entirely on surviving, but survive I would. Perhaps others felt like I did, but no one talked about those kinds of things. We all put on a brave face and carried on—except for the ones that kept quietly disappearing.

It wasn't just my class either. I wrote to my parents in late January about a 2/c cadet, or a junior at a normal college, who had been recommended for disenrollment due to conduct and subsequently voluntarily left on his own. The rumor was he'd had some kind of relationship with a female classmate of mine. In Chase Hall parlance, we referred to this as "the dark side" and he'd apparently been caught red-handed. On one of his last days, while many others from his class were ripping into us at a formation, he came up and told me casually that he'd stand in front of me for the duration of the abuse his classmates were doling out, thereby sparing me from their wrath. I genuinely appreciated his kindness, even if only for a few minutes, and felt bad for the guy, two years of his life lost and him now summarily kicked to the curb.

At the end of January, I was fortunate to make the roster for the summer's ocean racing program. I'd put my name in for the team weeks earlier, hearing stories about racing sailboats around New England for the bulk of the summer sailing season. After their first year, most cadets were sent out to the fleet to learn a little bit more about the "real Coast Guard," but a small group, pulled from the fall season's sailing team, were picked to take part in the ocean racing program. This was perhaps the best news I'd heard all year. All I knew at that point was that I'd been assigned to a boat named *Rampage* and more details would soon follow.

As the spring approached, I struggled through calculus and chemistry classes, hoping for little more than a passing grade so I wouldn't have to repeat them in the fall. Two things kept me focused. First, there was the approaching ocean racing season and in late winter we began to have meetings to prep for the summer. And second, Chuck had convinced me that if I went to stay with him over our spring break, there would be a plentiful supply of boats, surf, and girls. It did not take long for me to sign on for the trip, and I coordinated with my parents to ship a surfboard down to his parents' house in Port Aransas, Texas.

Spring break was in early March, and as promised, I spent a week on the beaches of South Texas. We also linked up with some other friends from the academy and took a road trip to South Padre Island for several days, where we immersed ourselves in the stereotypical spring break activities and managed to drive away a few days later without a criminal record. Chuck's parents had a boat as well, which we made good use of during the week. I managed to get in a bit of surfing and much of our time after the sun went down was directed at the third element of the spring break trifecta. I had met a remarkably pretty girl from his high school and by the end of the week she'd left quite the impression on me. As the week came to an end, Chuck seemed annoyingly excited to get back to the academy. He also blasted Creed songs on his stereo as we packed our bags to head back to New England, which was equally annoying. Me, sitting on the opposite end of the spectrum, had little to no desire to return to New London. Had it not been for my promise to myself not to quit and the ever-present Uniformed Code of Military Justice, I may very well not have returned.

By mid-March, I was back on the Thames, reorienting myself with the art of sailing a boat, and taking great care not to slip and fall into the frigid river. Early in the season, it was not uncommon for us to take brooms and sweep snow from the decks of the J-22s before casting off from the dock. By that point I'd purchased a fleece-lined sailing jacket for the summer and wrote to my parents about my concerns that I'd be kicked out before ever having the opportunity to really break it in. Having lost the privilege of civilian attire on our reporting-in day, there was a slight grey area when it came to sports uniforms. The jacket itself was nothing fancy, but to have something of my own that I could wear to practice, with colors straying from the light blue/dark blue uniforms (the jacket being royal blue) was a tremendous boost to my morale.

The spring sailing season didn't have as many weekend regattas, but I filled my on-the-water quota by volunteering to assist with some of the dinghy regattas that were held at the academy facilities. As April approached, I was spending Saturdays zooming around the river, moving buoys and coordinating starts and finishes for collegiate races. I had also found my place as a bowman on a three-person team with one of the better helmsmen on the varsity sailing team. How that happened was a bit of luck, but once again personality conflicts had given me an opportunity to step up and secure a spot with a crew that knew what they were doing. The bow work on a J/22 was not all that complicated, but the helmsman, a 2/c cadet, gave me no slack with a steep yet enjoyable learning curve.

By the first week of April, the class of 2003 had been granted 'Carry On' which meant, in simple terms, we were no longer required to run everywhere on campus for no apparent reason, we could look down at our food when eating, and didn't need to walk directly down the center of any and all passageways, nor did we need to make all of our turns in exact 90-degree angles. Walking like a normal human being around a corner was a real treat. The simple act of walking to class was an incredible feeling. The air had warmed up considerably from the depths of winter and the days were growing longer.

Comfortable now on the J/22s, I took each opportunity we had to push the little boats to new levels. The steady and bitter north wind in the spring would howl down the length of the river and I walked away from most practices with new bruises, scrapes, numb extremities, a runny nose, and a smile on my face. On a downwind leg, when loaded up, the bow of a J/22 would surge up and ahead as if about to come up on a plane like a speedboat, with a small bow wave breaking in front of it. While we never did get one to plane, the experience was a good introduction to pushing boats closer to their limits. An important lesson was to avoid exceeding that limit. Had those boats capsized in the early spring, we were not wearing anything remotely close to proper waterproof attire. Hypothermia was a real thing, and at times we were seconds away from being thrown into the river. All in good fun, of course.

The coaches zoomed around us in skiffs, at times yelling at us to do better and offering instructions. No doubt they would have quickly scooped us out of the water, but none of us wanted to take a chance in the near-freezing Thames. At the end of a downwind, with a brief reprieve from the wind, we'd hike out on the rail for the upwind leg, our faces numb in the stiff breeze as we tacked our way back up the river for another downwind run. If and when the bow hit a wave the right way, spray would soak us, and I learned to grit my teeth through the numbing cold. No one else seemed to complain so I'd be damned if I was going to.

When particularly nasty weather rolled through in the form of snow, rain, sleet, a gale, or any combination of the above, we spent afternoons prepping the larger boats for the summer ocean racing season. The early work consisted mostly of inventorying anything and everything that might be needed for the summer. Hours passed by as we pulled out sails, taking notes of which were in good shape, and which were torn or otherwise too far gone for another season. Each boat, *Rampage* included, had a large locker at the waterfront full of deck fittings, tackle, lines, sheets, chains, anchors, and every removeable piece of equipment that had been stored inside for the winter months. With a month of classes remaining, I knew that if I could squeak out another semester of passing grades, the summer was sure to bring new adventures.

Weeks before classes ended, there was a perceptible light at the end of the long tunnel I'd been stumbling through for the better part of a year. That was, right up until I got in trouble. Over the course of my first year, I'd become acquainted with what was a common occurrence; 'spirit missions.' Bordering on the edge of 'hazing,' there were innumerable instances where senior cadets would call on junior cadets to perform embarrassing acts or menial repetitive tasks in the name of fun and to build camaraderie. This idea is of course nothing new to the military. An argument could be made that 'Swab Summer' was a form of sanctioned hazing—like when I had to carry a bowling bowl around campus as punishment for looking around.

Within the culture that existed at the academy, there were times when unsanctioned events took place within Chase Hall, and I'd come to not think twice about them. Prior to my own trouble, I'd seen a Second-Class Cadet duct-taped to an office chair, wearing nothing but his underwear, pushed out into the hallway and left to fend for himself. One particularly strong Third-Class Cadet had put out a standing challenge that no other cadet could go for a period of two or three minutes in a room with him without screaming out in pain. While I'd never witnessed it first-hand, I'd heard the muffled thuds and bangs from outside his door, while a dozen or so of his friends cheered from seats in the upper bunks like spectators at a Roman Colosseum.

These things happened and over time, it became normalized. I have no opinion either way as to the worth of such antics, but in early April, 12 of us set out to celebrate Chuck's birthday late one night. The prank involved handcuffs, an electric razor, and edge-dressing, which was a thick tar-like paint we used to put a shine on the soles of our dress shoes. Unfortunately for us, someone thought it prudent to videotape the entire ordeal. The 13 of us (even though Chuck was carried out of his room to a nearby shower against his own will, he was deemed a willing participant) were charged with Class I offenses, which is a serious affair. I wrote to my parents that the official charge against me was *Abuse of Authority: Hazing* and that I was confused as to what authority it was that I had as a fourth-class cadet. My father, at this point with nearly 30 years of military service under his belt, wrote to me on the 18th that my statement "*is to be CONCISE*" and that I must throw myself on the mercy of the court. This was timeless advice that would serve me well throughout the entirety of my own career at any point (and there were several) where I had strayed from some regulation or the good graces of my superior officers.

The investigation dragged on for a few weeks, the 13 of us all in limbo while the paperwork made its way up the chain of command. First-Class Cadets wrote the initial report, then submitted their findings up to their respective company officers for review. I wrote to my parents on the 19th of April,

Twelve somber cadets, with black cloth bags over their heads, are led out into the mob as crowd control tries to keep them from ripping our ragged clothing off. Their arms are chained and connected to their legs with one mighty length of chain. As a path is cleared towards the steps up to the ropes, the angry crowd begins throwing rotten fruit. Some of the twelve fall to the ground. The black cloth bags are torn from their heads to reveal the bruises and scars from a month's worth of restriction. The crowd chants "crucify them, crucify them." Slowly, a man with many stripes on his shoulder boards steps up the ladder, and the crowd goes silent.

"4/c cadets, you have been sentenced to hang for the shaving of your friend, whose name will be kept secret so that he may continue a normal life after you have been disposed of. Do you have any last words?" We try to speak, but the screams of the crowd drown out our cries for mercy. One by one, the trap door releases, and the crowd erupts into cheers. The man with many stripes laughs.

Perhaps I took some artistic liberty with email, but it captured my mood perfectly. At the same time the Captain's Mast was hanging over my head, the summer sailing season was fast approaching and work on the boats had begun. By the time classes were nearly done, I was spending most of what little free time I had down at the waterfront, 'tinkering' on *Rampage* for lack of a better word. Boats are often referred to as holes into which the owner simply dumps money. The officers that oversaw us were in a unique position whereby they were not on the hook to fund the upkeep of the boats. That bill was paid by Uncle Sam as was the labor, chiefly in the form of a dozen cadets assigned to maintain each boat. The downside was that most of us had no clue what we were doing as we took things apart, cleaned them, and hopefully put them back together correctly. For reasons I've never understood, every branch of the military takes great pride in directing inexperienced junior personnel to take things apart and clean them before putting them back together, at times incorrectly. Most often, there was little need to do so in the first place, but it's a time-honored tradition. The magnitude of the endeavor to which I was about to embark on certainly didn't hit me on those early Spring afternoons. At sea, there is little room for error, and with each bit of safety wire we fastened to the rigging, we were unknowingly entirely reliant on our work to make it back to dry land in one piece.

On the day that classes ended in late April, I wrote to my dad about working on the bilge, replacing rotted wooden planks down below, and how I'd helped to re-attach a support beam using fiberglass up forward near the bow. During my high-school years, I'd learned a fair bit about fiberglass work as I was constantly fixing cracks and dings in my surfboards. The opportunity to do something hands-on was far more rewarding than the boredom of academia up the hill.

The academy waterfront had a substantial maintenance program as well with the essentials to run a small boat repair shop. I quickly learned that, in the absence of adult supervision, I was free to pick supplies and rigging to outfit the boat. There was a Chief Petty Officer who ran the waterfront, and he did not like cadets. At all. Looking back, I don't blame him. On several occasions he would catch us scurrying off with a handful of cleats or shackles and yell for us to bring them back. Unfortunately for his concerns over fiscal responsibility and stewardship, there were many of us and only one of him. By the time May came around, we'd pilfered most of the hardware, lines, and supplies that the waterfront had to offer. *Rampage* may have been old, but she was well-provisioned.

Along with the Chief, there were a dozen or so junior enlisted guys that worked on the boats. Electricians Mates, Boatswain's Mates, Damage Controlmen, and several other ratings were put to good use maintaining the fleet and fixing things that we broke. I would often shadow one for an afternoon to learn what exactly it was that they did. I took an immediate liking to them as most of the guys would spend whatever free time I had teaching me the raw mechanical skills of running a new fuel line or using a pneumatic sandblaster to clean up a rusted fitting. Over that summer, I found that whittling away at some random piece of corroded metal with the sandblaster did wonders for stress relief. I suspect that the guys knew exactly what I was doing, but they never seemed too bothered by it. Where others may have sought professional counseling, I was quite content to take out my frustrations by annihilating some poor piece of galvanized steel.

It was unintentional, but my interactions with the enlisted folks at the waterfront set me up well for my time in the Coast Guard. From those early days, I learned to carry a great deal of respect for the work that they did. It was true that the officers crafted a plan for the sailing season, but had it not been for the enlisted crew's expertise and hard work, none of that would have been possible. While I detested the textbooks, lectures, and unending tests up the hill, I took considerable pride in the work I'd accomplished at the end of each day down on the waterfront. There was something tangible about a day's worth of work there that I didn't experience in calculus tests or chemistry labs. The boats were slowly coming together, and we were increasingly ready to take on the Atlantic Ocean, or so I thought.

On the morning of May 12th, I finally got to my Mast, or trial as it would have been called in the civilian world. From what I can remember, the Lieutenant Commander that presided over the hearing focused his questions on the edge-dressing art that we'd drawn on Chuck. The first image was fairly straightforward, perhaps the most predictable thing 19-year-old guys would draw in an attempt to embarrass their friend. But the second thing we drew was more complicated and we obviously hadn't done a good job of it since he had no clue what it was in the video. With a straight face, he asked what the second image was supposed to represent.

I took a long breath. With me as my mast representative was Stan, who was the senior cadet on *Rampage* for the summer and would soon come to epitomize leadership in every sense of the word. Unfortunately, I was the only one in the room who knew what we'd tried unsuccessfully to draw. With the Lieutenant Commander looking at me from behind his desk, and Stan by my side, I focused and tried to keep my composure. Standing at attention, I said, "It was a Teenage Mutant Ninja Turtle, Sir," and then waited. Stan laughed briefly through his gritted teeth and then clenched his jaw even harder to stop himself. At the same time, the Lieutenant Commander looked straight down at his desk, covered his face, and I watched with cautious curiosity as his torso bounced with each uncontrollable bout of laughter.

I survived the mast with a whopping total of 60 demerits, 15 work hours, 15 marching tours, and 47 days of restriction once the fall semester began. But I would be free for the summer as there was no practical way to serve out my time while I was away from the academy sailing. I breathed a deep sigh of relief after being excused. Graduation for the class of 2000 came and went shortly thereafter. President Bill Clinton was the keynote speaker, and to this day I can't remember a thing he said. But after graduation, I was officially a third-class cadet.

Chapter 4

With classes over, our work began in earnest on *Rampage*. By racing standards she was an old boat, having been built about 20 years prior. Donated to the academy, *Rampage* was an old racehorse in quasi-retirement. She looked fast, but a deeper study revealed a tired hull and old rigging. By late May, we'd begun to find problems and, as is often the case with boats, we created additional problems on our own. The simple job of installing new compasses serves as a great example. Someone well above us in the chain (ie. an officer) had decided that *Rampage* needed new compasses, one for each side of the cockpit. Under sail and on open water, the helmsman relied on those compasses to maintain his course. At night, those two dimly lit compasses were even more important, as any surrounding clouds or landmarks were impossible to see as a reference for keeping one's bearings.

So in late May, our new compasses had already been installed and properly mounted into the deck, but a decision was made—again, likely by an officer—to reseal them with marine-grade putty in order to ensure the seal was watertight. A classmate of mine who had never removed or installed compasses on a boat was assigned to this task, but we had confidence in his abilities given that he ranked near the top of our class. He was a whiz at physics and was going to major in Electrical Engineering. Sealing some compasses was certainly well within his abilities, so we left him to it.

The rest of us worked down below to re-route a vent that directed air from the head, or bathroom, away from the cockpit so we wouldn't have to deal with the smell while sailing. As several of us were working in the cabin, we heard the rush of some fluid followed by swear words and the distinct smell of oil. What my classmate had done was remove the face of the compass instead of the whole unit which then spilled out the entirety of the very specialized oil in which the compass floated. The new compass was now essentially ruined, and we had one of many significant and self-induced setbacks.

The lesson I learned was that academic prowess did not necessarily translate into practical seamanship. The critical point to be made is the distinction between academic aptitude and hands-on application of those skills. How the compasses were fixed I can't be sure, but there's no doubt that the enlisted guys working at the waterfront were the ones who made things right again. At the time, as a cadet just entering his second year, I couldn't fully grasp their perspective, but I can appreciate it far better now. Many, if not most, of the enlisted folks at the waterfront had spent a tour or two underway on cutters. Many of them had also spent years at boat stations, serving not only as crewmembers but also as the ones tasked with maintaining the various small boats. Most of them had families at home and were far older than us. For them it must have been tough to see a dozen cadets, many of us still teenagers, destroying sailboats in the name of education. Making matters worse, in a matter of two or three years, we would be commissioned as officers and, with the stroke of a pen, hold a rank superior to theirs. I can't recall a single interaction with any of the petty officers where they showed any disdain for my presence. Looking back on that now, it was incredibly commendable on their part.

At the end of May, we took *Rampage* out on a shakedown cruise. The plan was to sail her out the Thames, into the Sound, then a short hop over to Block Island where we'd spend the night and sail back the following day. I was excited for another night away from the academy. Even with my fourth-class year in the past, it was still a welcome relief any time I was able to leave New London. We departed uneventfully one morning with a healthy breeze carrying us out past New London Light, then south and east past Fisher's Island. By the time we were on open water, the wind had begun to pick up considerably. If anyone had bothered to check the weather, it most certainly hadn't been me. We'd practiced sail changes on the river, at times switching a light jib out for a heavier one. Like any evolution on a sailboat, it was an exercise in teamwork, communication, and patience. On the river, in manageable winds, we seemed to be able to sort these things out without much trouble.

But as the winds picked up, we swapped one or two jibs out, and the process was slowed down as the bow dug into waves with cold sea spray pelting everyone on the deck. We reefed the mainsail, dropping it once, and then a second time to counter the increasing winds, which at this point were gusting near 40 knots. We were soaked, cold, and yelling, not so much out of frustration, but more so with the hopes of the wind not entirely drowning out our words back and forth. At some point, Stan called for us to rig the storm jib. As the weather deteriorated further, Stan seemed to move effortlessly around the pitching and rolling deck, offering subtle encouragement and guidance as we each tried to do our individual jobs. I didn't want to admit to anyone on the boat, Stan especially, that I had little idea what he was talking about. From down below, someone tossed up the smallest and most over-engineered jib I'd ever seen. It was intended less for generating power and more for stabilizing a boat in rough weather. We somehow had it rigged without anyone getting hurt, and the triangular blade of a sail did a good job of keeping the bow steady as we made slow progress towards Block Island.

Soaking wet, tired, and cold from the late-spring storm that was now giving us a good beating, we sailed on a reefed main and the storm jib for the remaining miles to Block Island. A narrow channel into the harbor knocked down much of the seas, but the winds still ripped across the Great Salt Pond, a large bay in the middle of the island. With our engine straining to make any headway, we navigated slowly under a dark late-afternoon sky towards our dock. Even with lines over, we were unable to pull *Rampage* entirely into her spot until we wrapped those lines around winches and muscled the boat into her berth for the night.

Afterwards, I was physically and emotionally exhausted in a way that I hadn't felt in Chase Hall. My body ached and my core was cold, but we had accomplished something remarkable, even if it had been dumb. In the grand scheme of open water sailing, it was a short hop from New London to Block Island, one that I would do dozens more times over the coming years. But on that afternoon, it felt significant, as if I'd circumnavigated the globe. It was strange given my sad soaking wet state, but I was quite content with our performance. We'd inadvertently tested ourselves and held up well against mother nature. Drying off as best we could, we wandered the town looking for a restaurant that might be open. The summer season had not yet begun, so our choices were limited until finally a bar took pity on us and served up hot bowls of beans and rice with salsa. We were a sorry looking yet incredibly appreciative group of amateur sailors.

Still not entirely dry or warm, but with full bellies, we fell asleep that night on *Rampage*, each of us collapsing on a canvas rack or a pile of sails, listening to the wind as it slowly died down through the night. By the morning, the storm had passed, and we sailed uneventfully back to New London. If anyone had been rattled, no one dared to show it. Stan had shown himself to be fearless. We had a Lieutenant with us, but a healthy distrust of officers had already been cemented in my mind, so as far as I was concerned, Stan was the *de facto* leader of our little misfit crew. I believe the first true lessons in leadership that I learned as a cadet came from that trip to Block Island.

As June neared, we were still busy preparing *Rampage* for the season to begin in earnest. Our first regatta was at Port Jefferson, New York. The little bit of information that I had from the older cadets centered around the hospitality extended to us by the yacht club, namely in the form of access to alcohol during and after the regatta parties. I, like most of my classmates, was younger than 21, but the snickers from the senior cadets provided clues of what was to come. The racecourse was a tight triangular route of big boats all trying to jockey for position and get ahead. We practiced hard in the week or two leading up to our departure, mostly focusing on our timing of tacks and jibes. Small mistakes would translate into big losses, those losses magnified even more so by a short racecourse. It would be my first opportunity to prove that I could keep up and I rehearsed the choreographed routine of letting one sheet off a winch, loading up the leeward one, and wrestling with a winch handle to trim it in the least amount of time. It was not difficult to perform those tasks in a static environment. At the dock, it would seem menial. But at sea, with the wind blowing, the ship rolling and pitching, and the crew all moving around, it would prove to be far more challenging.

From Port Jefferson onward, I spent the next few weeks at regattas each weekend, and we would normally get back to the academy late Sunday night or early on Monday mornings, equally tired from the sailing as we were from the post-regatta festivities. A typical week from that point on was a few chaotic days of fixing things we'd broken the weekend prior and squeezing in a few hours of practice in the afternoons. By Thursday, we were normally ready to go for an early departure Friday morning. Regattas, at least in and around New England, are a funny thing. At nearly each event, there was a fair mix of professional level racing sailboats, wealthy amateurs with their yachts and pampered crews, and then the hardcore racers whose ambition far exceeded the depth of their checking accounts. Friday nights were a social event where each boat was checked in and crews mingled and drank late into the night.

Trying to keep a low profile, we normally set off as a group with whatever contraband we could get our hands on and found a dimly lit dock or beach where we could sit and blow off some steam. Most often, we were well provisioned by the yacht club president or his appointed representative, who seemed to always take pity on our spartan cadet existence. Perhaps they felt it was their patriotic duty to facilitate a dozen or so future officers imbibing untold quantities of spiced rum. The officers with us were normally wholly consumed by the party and left us alone which created a low-threat environment. For me, those nights were a much-needed respite from the academy and provided the first opportunity in close to a year to relax. Moreover, I was feeling out the social hierarchy amongst the other senior cadets. I was a third-class cadet at that point, but still not entirely comfortable letting my guard down.

Inside Chase Hall, much of the pressure from the school year had subsided. Most of the Corps of Cadets were away for the summer, either at internships or aboard cutters. This left us free to do as we pleased during the nights in Chase Hall. While most days ended with us tired and sunburned, we found a few opportunities to have some fun. At some point, I had purchased an inflatable children's pool that fit perfectly in the shower area of a bathroom near our rooms. With all four shower heads pumping hot water into it, an hour or so would pass until it was a quasi-usable jacuzzi. With Chase Hall empty, we were able to get away with our amateur day spa for a week or two until someone caught on and we were directed to remove it. Rather than dispose of the pool, Chuck and I spent a morning setting it up in the room occupied by two second-class cadets on the ocean racing team. We spent several hours running back and forth with pitchers of water, filling it as high as we could. To further complicate any attempt to remove it from their room, we lathered up the inflatable sides with shampoo and soap. I'm not entirely sure how they got that pool out of their room without flooding the adjacent hallway, but somehow they did. They would return the favor later by waxing and buffing a centerfold cutout from Playboy onto the floor in my room, which necessitated hours of painstakingly scraping it off my floor with the bayonet from my rifle. There was no doubt more trouble than that, but we stayed under the radar of the few officers who made occasional rounds to sniff out their errant children having too good of a time.

Our capstone event would be the Newport-Bermuda race towards the end of June, where we would race from Newport to Bermuda, spend a week with some smaller races in and around the island, then sail *Rampage* back to New London to close out the ocean racing season. In preparation for the regatta, we'd spend close to a week in Newport making final adjustments to the boat. As if that wasn't enough of a good deal, Chuck had arranged for his mom to bring some of his friends up from Texas for the week, and she had rented a small apartment for us all so that we didn't have to sleep on the boat.

Of particular interest to me was the fact that Leslie, who I'd met over Spring Break, was coming along. That week in south Texas had been quite an event for me. It was my first experience with country music and, having taken a serious liking to that tall and skinny brunette, it was as if many of the honky-tonk overly twangy pop country songs blasting from speakers in the beds of pickup trucks were written with my most immediate feelings for her in mind. I had left Texas uncertain if I'd ever see Leslie again, so as our week in Newport approached, I was as excited to see her as I was to sail to Bermuda.

Newport is a picturesque sailing town, and as we tied up along the waterfront, I was drawn to its New England charm. Quaint, beachy, and ungodly expensive, it remains one of my favorite towns in America. The week we spent there provided enough downtime for us to familiarize ourselves with the town and for me, it was the first opportunity to wander around a bit on my own. We hung out at the rented apartment for a few days before Chuck's friends and family came into town. One of those afternoons still stands out. I had been in the Coast Guard for nearly a year at that point, and most of my decisions were either made for me or heavily influenced by someone above me. Taking an inventory of what I'd brought, I realized that I'd forgotten nail clippers. During swab summer and much of the academic year, there was a heavy price to pay when one forgot anything. Some sort of physical punishment or ridicule usually followed, no doubt an attempt to drive home the importance of being prepared for whatever might lie ahead. But as my friends lounged around on couches, I quietly slipped out and wandered downtown, taking a few shortcuts that turned into longer detours until I found a pharmacy. With nothing planned for the rest of the day, I took an even more circuitous route back to the apartment and took a great deal of pleasure in knowing that no one had even the slightest concern that I'd been gone. For me, that sense of personal freedom to simply wander about aimlessly for an hour or two was remarkable and remains a moment that I still cherish.

Chuck's mom and his friends, including Leslie, showed up a day or two later. For the next few days, my attention was exclusively focused on her. It didn't take long for the two of us to pick back up where we'd left off. There was an easiness in our conversations, the kind of talking that goes on far longer than either of us realized and kept us up until last call. And as a cadet in downtown Newport, I was quickly burning through the small amount of money I'd been able to save up over the past year. It didn't help that each night we were running around town until the early hours of the morning, dollar bills literally falling out of my pockets as we all stumbled from one establishment to another. As the start of the Newport–Bermuda race approached, I ran into two problems that have plagued sailors for centuries, the first of which has ensured an endless supply of able-bodied seamen willing to embark on a voyage since the dawn of time. I was broke. The second problem was wanting to stay in Newport for as long as I could with Leslie. With that, I'd also unknowingly joined a long and distinguished line of sad sailors who'd faced the same conundrum. The decision is often made for the sailor, not by the sailor: once all the money (and then some) had been spent, that poor hapless soul finds himself aboard ship, resigned to sober up and recoup his losses with his sweetheart left behind—at times on less-than cordial terms—as a ship sets sail for the open sea.

Despite being entirely lovestruck, I dutifully said goodbye, with an agreed upon plan in place to see her again later that summer, and boarded *Rampage* for the start of the race. It was a rather uneventful start to such a large regatta. When sailing around a smaller course of buoys, the start was usually one of the more intense evolutions with each boat jockeying hard for the optimal position. However, with hundreds of miles between us and Bermuda, it was more relaxed and as we crossed the starting line and trimmed sails, we each settled in as best we could for a week at sea. As dusk approached, I looked back behind us, and the lights of Newport disappeared below the horizon. My mind wandered back to that morning with Leslie. I shifted my gaze forward again, toward the dark mass of ocean and sky in front of us and took long breaths of the salt air until the lingering thoughts of land left my mind.

We'd only done one race up until that point that was a multi-day event, so I had little appreciation for the sleep deprivation awaiting all of us. Split into three groups, or watches, two would be up and about sailing the boat for anywhere from four to six hours at a time. There was obviously some overlap between the groups, and at times all of us were required to change out sails or make temporary repairs to something that had broken. On a sailboat, out on the open ocean, a sailor finds his downtime centered around two things: food and sleep. A distant third is a mostly foolish attempt to dry oneself off and change into the least damp clothes on hand. This foolish goal is most often in vain as the first wave over the bow soaks you once again to ensure that you'll spend the remaining hours on watch wet and cold.

We had been issued foul weather clothing which consisted of heavily used Gore-Tex jackets and pants that offered some protection from the elements. But our undergarments and footwear were our own responsibility. I had a pair of neoprene zippered boots that were good for sailing dinghies around a racecourse for an afternoon, but they were far from waterproof. Regardless of how many times I changed my socks, my feet were cold and wet for the first days of the regatta, so cold that I was convinced both my feet were going to fall off. I learned quickly that this was the limiting factor in my offshore ensemble and warming them up in between watches was a painful affair. Underneath the Gore-Tex, I layered what little fleece I had with me and tried not to dwell too much on the constant discomfort of being cold, wet, and tired.

Those first few days on the Atlantic were my introduction to true sailing. We ate food that wasn't very good, drank lukewarm water from gallon jugs, slept on rolled-up wet sails, and tried our best to keep our spirits high. Stan kept us focused on the race. Day or night, rain or shine, his positive attitude and gentle nudges kept us all in line. While the helmsman worked the tiller against the wind and waves, the rest of us sat perched on the windward rail, eating waves and spray as *Rampage* plowed southeast towards Bermuda. If left unchecked, one's mind wanders to some strange places at times like those. Whether it was chills or hunger or fatigue, something always seemed to snap me back to the present moment where I'd blink a time or two and stare off at the distant grey horizon to regain my bearings.

We hit the Gulf Stream late one afternoon. This, naturally, coincided with a thunderstorm. What had been cold waves of water breaking over the bow turned warm almost instantaneously. I was hiked out on the rail with several others, chills running down my spine and within a few minutes, we were laughing at the warm water that pelted us and ran down our necks, soaking our undergarments. Far out into the Atlantic Ocean, and still early in the race, we may have been tired and wet, but the absence of the cold was a tremendous morale boost for us. The storm around us persisted for some time into the night, lightning strikes at times coming close to our 40-foot boat, but if there was anything to be worried about, I was blissfully unaware. At least I wasn't cold anymore.

With each passing watch, I tried to wipe myself down with a towel and some fresh water once down below. By the second day, I, along with everyone else on *Rampage*, had given up on such trivial pursuits. Regardless of how good of a scrub one gave themselves, it would only be a matter of minutes up on deck during the next watch when the first wave of water would soak us, and we'd be once again coated in a thin film of salt for the remaining hours on watch.

Food and sleep thus became the sole concerns during my off-hours. No time was wasted on plates or napkins, and each shift ended with a mad scramble to consume as much food as was available before climbing into a rack and quickly falling asleep. *Rampage* had an exposed engine down below that we would run for periods of time to keep the batteries charged. Four of the canvas racks were directly next to the engine which made sleep challenging over the noise, but I learned that sleep deprivation has its benefits, namely the ability to nod off nearly anywhere. By the third day, I'd traded most of my foul weather gear for board shorts and a t-shirt as we were more than half-way to Bermuda, and it seemed as if we'd turned a page. Both the water and air temperature were comfortable and, at times, almost too much, with little more than a light tropical breeze across our bow to keep us cool. We were close, but still days away. As another day passed, we all strained towards the horizon, looking and hoping for that first craggy rock to appear in the distance.

Chapter 5

I first caught a glimpse of the island in early daylight, thoroughly confused at what I saw. Bermuda was supposed to be lush and tropical, yet as a faint object on the distant horizon, it looked more like a rock sticking out of the ocean. Having formed 30-million years ago, the archipelago resulted from a momentary disturbance that produced a solitary volcano in the middle of the Atlantic Ocean. That incident yielded almost 200 smaller islands first discovered by the Spanish in the 16th century, then settled by the English in the early 17th century.

Bermuda had, in past centuries, been known as the 'Isle of Devils,' dangerous not only due to evil spirits, but also from the Atlantic storms that ravaged its rocky coast. Fortunately for us in the 21st century, the evil spirits had long ago been chased away and modern meteorological services forecast mild weather for the entirety of the regatta. I stared at that first solitary rock that appeared over the horizon for hours during my watch. At a speed of six or eight knots, the time passed painfully slowly. Worse yet was that the finish line was farther than the entrance to the harbor, meaning we would have to sail around a more circuitous route before finally heading into the Great Sound. This added maybe an hour or so to our race, but after the better part of a week at sea, it was torture.

Our race ended with a blast of the air horn from a signal boat, and we proceeded into the harbor to tie up. Stepping onto solid ground is always a strange feeling after any time spent at sea. But once ashore, there was too much to take in and whatever fatigue I felt disappeared. I was in sensory overload as we located luggage that we'd shipped in advance of the race and found our way to showers at the yacht club. Washing off a week's worth of salt, sunblock, and crud is no easy task, and it took multiple attempts at lathering up before I felt anything close to being clean again. Finally, wearing fresh clothes, we were tanned, tired, and clean, but not one of us had sleep on our minds.

Our safety officers had given clear instructions on what we could and couldn't do during our liberty hours, which would be most of the upcoming week before sailing home. Alcohol and the plentiful rental scooters around the island were expressly forbidden. We nodded our heads in understanding, and then once out of sight of the officers, found our way to the nearest scooter rental shop. We explored the island for the remainder of the day before parking our scooters and finding an appropriate establishment from which to disobey the other general order. With a legal drinking age of 18, we quickly found ourselves celebrating our success in the regatta in the most appropriate manner expected from any good sailor.

Of particular importance to the island and the regatta was Gosling's Black Seal rum. One drink, called a Dark 'n' Stormy was the *de facto* official cocktail of Bermuda, so much so that the Gosling's store provided prizes and ceremonial titles to anyone who successfully partook in drinking Dark 'n' Stormy's at any of the officially licensed establishments on the island. We were intrigued. For tracking purposes, Gosling's provided us with a small card that would be marked at each bar we visited and successfully finished our drink. My efforts during the following week were rewarded with an honorary promotion to the rank of Captain, something I was unfortunately never quite able to pull off while serving in the Coast Guard.

There was a day or two of sailing that week where we raced short courses around buoys, but for most of the week, there was little to interrupt my first taste of liberty in a foreign country. I was up each morning with the sun, mostly because I slept on the deck rather than down below. Most nights, we weren't back to *Rampage* until well after midnight. I'd pull a cushion or blanket with me up on the deck next to the cockpit and fall asleep, only to wake a few hours later and repeat the previous day's activities. We'd take our scooters all over the island to beaches and random points of interest, sometimes staying for hours, other times only for minutes before losing interest and taking off in another direction. At least two of the scooters sustained damage, one of which was due to someone's attempted jump over a dirt berm.

As the sun set to the west late each evening, we'd park the scooters and transition to our other activity, finding another sanctioned bar or pub to drink yet another Dark 'n' Stormy. Within the first three days, my paltry checking account was near zero, and we still had several days to go. I called my parents collect from a pay phone and recall telling my mom I needed money. My dad, to his credit, mumbled something in the background about how I'd finally become a sailor and that my mom shouldn't ask too many questions. Within hours, my mom had deposited a few hundred dollars in my checking account, enough to get me through the remaining days.

That week was a highlight of my cadet experience. Far enough away from New London to forget about the academy, we ran at full speed, day and night, sleeping for only a few hours each night before starting again. By the time we left, we were even more sunburned, tired, and incredibly happy to have had the experience. Ahead of us, the sail back home would be just as arduous as the race had been, albeit at a slower pace.

Reluctantly leaving our tropical paradise, we hit bad weather within the first two days and at some point the rubber gasket that surrounded our rudder post came apart. This created a gap where there was no longer a watertight seal, and we quickly had saltwater spraying up and into the cabin. Within the first hour, a foot of water sloshed around down below and one of us, for the next few days, was constantly down below with a thirsty-mate pump and a five-gallon bucket. We'd pump it nearly full of water before handing it up to someone on deck who would then throw it overboard. For at least a day, we were in heavy seas, and each time we opened the hatch to pass the bucket, a passing wave seemed to dump as much water down the hatch as we'd managed to throw overboard.

Here Stan's leadership once again shined. It was nighttime, I was soaking wet, up to my shins in water, and pumping as quickly as I could to slow down the flooding. At the aft end of the boat and behind the engine, I could see water spraying up each time the stern came down from a passing wave. My legs were pressed against a bulkhead to hold myself steady as *Rampage* fought her way through the storm. Occasional lightning would cast an eerie backlit shadow over the hatch cover as the rest of the crew worked the lines and sails up on deck. At some point I laughed, wondering what my mom would think of our present situation.

Then Stan slid the hatch open and peered down, nearly upside down, and his smile was the same as it would be on a calm sunny afternoon on the river. He asked, "How's it going?" Of course, Stan knew the answer, but he was checking up on me nonetheless. If Stan wasn't scared, then there was no reason for me to be either. I passed him the bucket, a wave of water came through the hatch and soaked me once again before Stan handed the bucket back down to me and went back to work on the pitching and rolling deck. In my first full year of being a cadet, there had been no lesson in leadership compared to what I'd just witnessed.

Our safety officer, the only 'adult' on the boat, had fallen on deck a day or two earlier and seriously hurt his back. After that, he had laid himself up in a rack down below and offered some guidance here and there, but for the most part Stan was running the show. For the remaining days at sea, *Rampage* was in a state of sort-of-sinking and the officer in charge of us was out of commission (no pun intended). We were, in a sense, on our own to make it back to the academy. The weather at some point improved and the flooding subsided a bit as the seas calmed, but we continued to pump water out of the bilges for days on end. A day or two later, we again encountered thick fog and hoped that our radar reflector, a round metallic object similar to a disco ball that we'd rigged near the top of the backstay, would provide enough of a radar signature for any lurking container ships to see us and hopefully avoid running over *Rampage*.

After about a five-day transit, we made it back to the mouth of the Thames and, lowering the sails, we unceremoniously motored up the river and tied her up. We'd started the trip back tired—which was entirely our fault—and now found ourselves even more so, once again covered in a crust of salt and grime, our eyes stinging as we walked our way back up to Chase Hall. I was, for perhaps the first time, glad to be back there. It wasn't flooding, my bed wasn't damp, and I had plenty of clothes that weren't soaked twice over in salt water and diesel. We would stay at the academy for another night or two before heading off towards New York City to meet the academy's training ship, *Eagle*, where we would then embark for four weeks of training.

Most cadets during their second summer spend five weeks aboard *Eagle*, but our time was cut short by the timing of the

Newport-Bermuda race. For this, I was incredibly thankful. The benefits of *Eagle* were dry racks, hot food, and far less likelihood of sinking. The downside to *Eagle* was that it was much like the academy in that our days were planned out for us and very structured. It felt strange to be back in an environment where we were told what to do and when to do it rather than making those decisions on our own as we had done onboard *Rampage*. I had taken for granted the freedom afforded to us while sailing her. Only a week ago, five or six of us on deck would have been responsible for navigating, sailing, and steering the boat, but on *Eagle* we were relegated to roles where six of us would line up and wait to be told to haul on a line, then yelled at again to stop. Much had been said about *Eagle* being a capstone leadership experience for us, but as I stood holding a line and staring at the back of one of my classmates, both of us waiting to be told when to pull, I saw few immediate lessons in leadership.

The other thing that I saw, and one that I learned quickly from, was the near-daily ritual of cadets getting in trouble. Whether it be shenanigans on the boat, or underage drinking in port, there was always someone in trouble. I met up with *Eagle* in New York City for the 4th of July. The pier was standing room only for the fireworks, and most of us easily found one or two pretty girls to invite aboard for the fireworks. For many of us, this was a 'fish in a barrel' moment. As the fireworks ended, the girl I was with asked if I wanted to go to a bar with her and at this point, I realized she was 21. At 19 myself, this was a risky proposition. In uniform, there was a high likelihood that I'd be served with no problem, but I'd already seen too many of my classmates face restrictions on the boat for doing just the same. With great reluctance, I declined her offer.

Kicking myself for the rest of the night, I realized the wisdom of my choice the next day as I watched yet another classmate stand in front of the Commanding Officer as his punishment for the previous night was doled out. I wrote to my parents in mid-July, during a port call in New London, that *"Some of the kids who got alcohol violations had their parents on board today and yesterday, and I kept thinking that if that was me and all I could do was see you onboard for a few hours, I would feel like shit."*

While in New York City, my parents had come to meet up with me and the thought of messing that up made me sick to my stomach. The lesson I took away was one of risks versus rewards. Yes, it may have been nice to have a beer with a pretty girl, but not at the expense of spending the rest of my summer restricted to the boat or worse. There were far too many of us crammed together not only with each other, but also the senior cadets, officers, and crew of *Eagle* that were constantly watching for us to err in judgment. I laid low and did my time, thankful when our at-sea training period was complete.

After *Eagle*, I had about three weeks off for a summer break. I had worked hard to coordinate a trip for Chuck to come back to Virginia Beach. My ulterior motive was, of course, to grease the skids for Leslie to come up as well, and it worked. The three of us hung out for a week or two, most days spent at the beach enjoying the easy days. I was entirely obsessed with her at this point, and anyone seeing the two of us together would have no doubts about that. We were officially dating.

As the end of August approached, I was back in New London, settling in for the beginning of my third-class year. This was, by most accounts, an academically challenging year where one was no longer harassed by cadre and left alone to focus on their studies. For many, this was a relief. For me, I was apprehensive. We were taking close to 20 credits and, with the academy being an engineering school, the course load was skewed heavily away from the arts.

Having survived both Pre-Calculus and Calculus I, Calculus II would be the next and hopefully final challenge in the math arena. I had barely survived Chemistry I and II, but now Physics I was on the docket, and I was none too thrilled. Rounding out the technical classes, I had Nautical Science to contend with. Sailing had given me a leg up on the realities of seamanship. This would prove to be a double-edged sword. At sea, plotting a position on a wet chart, under a dim red light, with the rolling seas was as much art as it was science. There was understandably some wiggle room. In the sterile environment of a classroom where time was not working against you, the instructors insisted on perfection, deducting points for the most miniscule errors.

On a brighter note, the sailing schedule had been released and I would spend the fall season once again aboard *Rampage*. It kicked off in late August and there were a few overnight trips, either spent aboard the boat or down at the Naval Academy. I wrote to my parents of both the Shields Cup and the McMillan Cup, knowing little of either, only that, in my own words, *the McMillan Cup was a premiere event in college sailing* and that I had slim to no chance of making the small team that would travel down to Annapolis to compete.

I was also serving out my restriction from the hazing incident at the end of my fourth-class year. Fortunately for me, I was tied up with regattas on most weekends, so I missed the monotony of being restricted to Chase Hall for most Saturdays and Sundays. There were, however, times when I was required to attend 'Restricted Cadet Formations,' usually on Friday nights or Sunday afternoons once I was back from sailing. In addition,

I had to squeeze in hours of community service and marching tours, which added to the already stressful third-class academics.

I was frustrated with the academic rigors of the academy. Joking with my parents, I wrote on the 31st of August, 2000, that *I will not need to differentiate the area under two curves rotated around the y-axis when I am on a cutter. I think that I am gonna wakeboard behind my cutter. And there will be surf calls in every tropical deployment.* Clearly, I was still operating at least partly under the false assumptions that I had made while watching *Baywatch* reruns during my senior year of high school.

I was also receiving letters from Leslie and making plans to visit her during the long weekend over Veteran's Day when I would finally be off restriction. These letters offered a glimpse of better times to come and something to look forward to. As I pressed on with the fall academics, I got word from my advisor that I'd received a 2.08 GPA for the spring semester of my fourth-class year. The classes that I had potential to do well in, namely Literature, Morals and Ethics, and Nautical Science were graded at a B-, B-, and C- respectively. Had it not been for the 20-plus credits each semester, I likely would have done much better in the arts, but my time was spread unequally among the classes and the lower grades in the above three were a casualty of my nascent time management skills. While nothing to brag about, my GPA kept me above the 2.00 cutoff that had already forced many others out the door.

As September arrived, the temperatures began to drop, and our sailing season was in full swing. I wrote home to my parents about one of our boats, *Eddystone Light,* that had hit a buoy while rounding it during a race after having misjudged the currents that rip around Fishers Island. Jon, who would become a very good friend of mine over the coming years, had taken a chance, and it hadn't played out as he'd hoped. On *Rampage*, we had kept farther off from the buoy after seeing Jon's misfortune.

Years later in the Coast Guard, I would come to see that oftentimes the organization operates with a zero-defect mentality, by which anyone who makes an error, large or small, is likely to face career repercussions. Among the officer corps, these incidents are often recorded in one's annual evaluations. Promotions to higher rank are competitive and one error in judgment, whether it be an honest mistake, a calculated risk, or simply a screw-up, could mark the end of one's career.

The collision between *Eddystone Light* and the buoy did cause some damage to the boat, but there can be little doubt that for the cadets who witnessed it that day, we learned an invaluable lesson in seamanship. While I'd received a C- in Nautical Science I up the hill, I was fortunate to have the benefit of practical lessons on the water on a weekly, if not daily, basis. This was once again an example of the dynamic environment on the water that simply could not be duplicated in a classroom. I could very easily tune out a lecture in a classroom. I couldn't do the same out on the water.

There is no lecture or textbook that can convey the subtleties of reading the wind and waves, feeling what the currents are doing to one's boat, and making immediate decisions on how to best navigate around the many hazards in and around Long Island Sound. Whether on the Sound or the river, these lessons repeated themselves. The engine on *Rampage* failed to start back up one afternoon on the river and we were forced to take her back to the dock under sail power alone. This was an easy task for a smaller boat, like the FJs or J/22s, but a boat like *Rampage* took quite a bit more finesse to work her in between the docks and finally tie her up without damage. The ever-present element of teamwork was there, but more importantly we were forced into a situation where we had to quickly alter the normal routine. We were simultaneously trying to think ahead and act reflexively. Take the sails down too early and *Rampage* would lose momentum, drifting down into another boat or another pier. Leave the sails up for too long and she'd prove unwieldy once dockside and likely be damaged. I wrote to my parents about how anytime I found a comfortable place to sit idly and take in the salt and sun, Stan Hudson was on me to get back up and get to work; there was always something to do and always something to learn. When I was on the water, I was learning practical lessons that would stay with me for the remainder of my career.

Up the hill, the fall semester marked the start of yet another season of scandals. A first-class cadet, or senior, was leaving after having been caught up in some sort of salacious affair on *Eagle* with one of my classmates. While aboard, some had embraced *Eagle*'s nickname of the 'love boat' or 'dirty bird,' and unfortunately gave in to lustful temptations. But the bigger scandal was rumored to be cheating by some of my classmates on their PQS, or Personal Qualification Standards. This was a dictionary-sized booklet we'd each been given before the summer and were told to complete by a certain date. It was brutally dry subject matter that required us to delve deep into Coast Guard Instruction Manuals and operating guides to answer literally hundreds, if not thousands, of questions related to all manners of Coast Guard operations.

Most of it was ship-centric and dealt with the routine and emergency operations aboard a cutter. I had accomplished exactly none of it during ocean racing but had buckled down for my four weeks aboard *Eagle* to plug away at it each night. While many of my classmates lounged around or stirred up trouble during their downtime, I spent most nights in the small ship's library slowly picking away at my PQS. I returned in the fall with most of it complete. But apparently some from my class had opted to share their answers among themselves and turn the work in as their own. The academy preaches integrity, so incidents like this were not taken lightly.

After first hearing about it, I reflected on my choice to buckle down in the library each night aboard *Eagle*. While my academic performance was sub-par, I had very early on made a conscious decision to keep my integrity as intact as I could. During my fourth-class year, I heard some rumors of classmates who were storing formulas on their calculators that would help them with calculus exams. Two things occurred to me at that point, the first being that this would likely be seen as cheating if they were caught, and the second being that I doubted I was even smart enough to figure out how to load and save a formula onto my calculator in the first place. In that sense, I was too dumb to figure out how to cheat. But I had earned my 2.08 GPA during the previous semester and if one day I was fortunate enough to walk across a stage as a graduate and newly commissioned officer, my no-doubt low GPA would be no one's but my own and I would hold my head a little bit higher because of that.

Chapter 6

The offshore sailing coach emailed me in the afternoon on September 13th, 2000, to tell me that I'd been selected as the alternate crewmember for the Navy 44 team. I'd be travelling down to the Naval Academy for both the Shields Trophy and the McMillan Cup. There were nine of us hand-picked from both the dinghy and offshore teams, and we'd begin practicing as a team in mid-September. I quickly wrote to my parents, telling them that it was the best news I'd gotten in quite some time.

It mattered because there were cadets senior to me with more experience yet somehow I'd been picked over them. Since joining the sailing team, I'd worked tirelessly at proving that I could hold my own and this was tangible proof, a validation that my efforts had not gone unnoticed. As an alternate, I'd likely be relegated to riding out the regatta on the coach boat, but for a kid who'd walked onto the sailing team with no experience, I was happy just to get the invite. Someone thought that I had something to contribute. Moreover, my last weekend of restriction would be spent down in Annapolis at the McMillan Cup, where I'd spend days on the water and nights free to wander around the town.

In addition to prep work for the races in Annapolis, I was still practicing with the regular offshore team and struggling to keep my head above water with academics. Physics was by far the most challenging class. In Calculus II, I had somehow managed a 71 on my first exam and reported to my parents that this marked the first time in three semesters that I had passed the first exam of a Calculus class. I had learned by this point to take comfort in the smallest of victories. Stan Hudson helped me study for Physics, and I managed to pass the first exam in that class as well.

It didn't dawn on me at the time, but the rigors of the academy were helping to develop a trait that would benefit me for the rest of my life. I was busy during almost every waking hour of the day, but in many ways, my tone in emails to my parents indicated that I had become rather content with it. One of the sailing coaches worked out a deal where I would serve my remaining work-hours helping run a regatta down at Jacob's Rock. This arrangement meant I'd be wearing a pair of shorts and a polo shirt rather than a formal uniform. It was also likely that I'd be barefoot and driving a small boat in the midday sun, which meant that serving out my last few hours would be a remarkably pleasant experience. I had all but shed any longing to idle away an afternoon. If my time was spent making progress towards some goal, whether it be academics or sailing, I had a purpose and I was happy.

Well into my second fall season, the Thames continued to provide lessons in seamanship. Normally, we sailed a familiar route around the river just east of the academy grounds, with Jacob's Rock almost always in sight. However, in late September, we had an officer filling in as the coach for a few days while our normal coach was away competing in a regatta himself. He devised a plan where we sailed further north up the river than any of us had been before, rounding a buoy that none of us knew anything about. *Rampage* drew nearly seven feet of water, and as we tacked west out of the channel, we quickly found ourselves in slightly less than seven feet of water, the keel no doubt dragging itself along the muddy bottom.

Making matters worse was nearly four knots of current that was working against our efforts to get back towards the channel. What had started out as a straightforward day of practice on the river turned into a scramble to keep *Rampage* making some way back towards deeper water. Running hard aground is generally frowned upon within the seagoing community. It is never a good strategy to rely on luck to remove oneself from a sticky situation, and yet here we were, blindly hoping that the keel didn't find some unmarked Volkswagen-sized boulder submerged on the river bottom. There must have been some significant wind that day as well, as we broached twice that afternoon. I wrote to my parents about the experience with the shallow flats, a new helmsman struggling to feel the boat out, and how being in trouble that far up a river reminded me of the movie *Apocalypse Now*. There were, at times, rewards for pushing oneself, but that day served to remind us all that one must walk a fine line or risk paying the price for poor decisions.

By mid-October, I was nearly done with my period of restriction from the previous spring. I'd completed the last of my work-hours down at the waterfront helping run regattas and had marched around the quad with my rifle for hours at a time to burn down my marching tours. For my first weekend of liberty at the academy, my wild plans consisted of going to see a movie and maybe buying some new running shoes at the mall.

The punishment doled out to me was, in the grand scheme of academy discipline, a fairly stiff sentence, but sailing had made it tolerable as I'd spent most weekends at least partially removed from Chase Hall and somewhere on the water, whether it be the Thames River or further out in the Sound. If the purpose of restriction, work-hours, and marching tours was to dissuade cadets from making poor decisions again, it had certainly worked for me. There was little in the form of rehabilitation, with the punishment being, I believe, intended solely as a deterrent from future misconduct. Many of my classmates, especially those involved with the hazing incident, would find themselves in a continuous loop of trouble for years to come. I believe most of my cohorts from that evening were eventually expelled for misconduct or grades. Thankfully, I had recognized that I was prone to making mistakes, but after the fact, I committed to never making them again. I would go the rest of my cadet career without any other major disciplinary incidents.

As October rolled on, we were busy prepping for a trip to Annapolis to compete in the McMillan Cup. Just prior to our departure, my academy advisor emailed me with my mid-term grades, which were abysmal. I was, at that point, failing both Calculus and Nautical Science. Both Physics and Macroeconomics were a D and Government was a C+. I wrote to my parents that despite all of that bad news, I was confident that I was doing well in tennis. As if that wasn't bad enough, the few of us travelling to Annapolis were instructed to bring our issued sport-coats and pleated khaki pants for some kind of reception that would be followed by yet another reception at some high-ranking official's residence. The Commandant of Cadets, a captain, was an avid fan of sailing and had taken it upon himself to plan some events for us while we were in Annapolis.

The absolute last thing that a handful of cadets want to do on a weekend away from the academy is to be around the senior leadership from the Coast Guard Academy or any other academy. Yet, here we were, our liberty hours reduced to only a few, and forced to endure an evening of dry conversation with senior officers from both the Coast Guard and the Navy. I'm not certain which was more depressing for me—my grades, or my social life being planned out for us in Annapolis. I reluctantly spent an evening in ill-fitting pleated khaki pants and a blue blazer, and it occurred to me that there was perhaps a bit of prestige to be involved in the regatta. The McMillan Cup remains the oldest collegiate sailing event, conceived in 1930 as an intercollegiate large yacht racing event for the New England and Middle Atlantic states. In the early years, it was dominated by Princeton, Cornell, Harvard, Dartmouth, and Yale. Through the decades, the maritime colleges began to replace the Ivy League schools as winners of the cup. In 1950, the Naval Academy had offered to host it, as their waterfront offered the best fleet of large sailboats, and the regatta has been held there ever since. Schools were invited from around the country to race and, to this day, it still provides a unique opportunity for colleges to compete with large boats.

As expected, I was the alternate for the regatta and spent both days on the coach's boat. While this was far from being in the thick of the racing, I kept an ear out for the little nuggets of wisdom that our coach would pass along as he watched our team maneuver around the course. I was keenly aware that my attitude had played a part in securing a spot on the team, so even though I wasn't competing, I was happy just to be there. At this point, I was quite capable at individual tasks like trimming a jib or a mainsail, but the art of tactics was still very new to me so I gleaned as much as I could from my seat on the deck of the coach's boat. St. Mary's College won that year, and I had little to say to my parents in an email after the weekend other than I had offered to drive the van back with the rest of my team passed out in the back.

Academics caught up with me within a week or so of returning from the Naval Academy. I had managed to bring my grade in Nautical Science up to a D, but my academic advisor was pressing me still. She had me forward my grades to my company officer, who was a Lieutenant. There were eight companies, each nominally led by a Lieutenant or Lieutenant Commander. As cadets, our interactions with them were few and far between. For the most part, they were middle management, their offices located in Chase Hall, where they could, in theory, keep an eye on us. Part counselor and part prison warden, they existed to set an example of how we'd be expected to conduct ourselves in the future.

I can't say that every cadet felt the same way I did, but I had grown distrustful of officers in general. In New London, most interactions with them were punitive, directly or indirectly. The academy was just as hard as I'd expected and the officers who oversaw our day-to-day lives were an easy target for my frustration. Between my academic advisor and my company officer, one of them had emailed my sailing coach, and when I went to practice the following day, he told me not to come down to the waterfront until my grades improved. I wrote to my parents in late October that he had said something to the effect of me being *no good to him if I fail out of here.*

At some point during my fourth-class year, an advisor had asked me if I knew what a self-fulfilling prophecy was. I replied that I did not, and he explained in rather simple terms that if you believe something will happen, one might begin to take steps to actually make it come true. He was asking me this after reviewing some of my grades and then listening to me justify my lackluster performance with excuses about how I was not very good at anything that involved a calculator. Typical of a 19-year-old, I had dismissed his attempt to steer me towards a more productive outlook on my cadet career.

This was a critical piece of advice, and I missed it entirely. It's likely that had I even managed to absorb the lesson, I was ill-equipped to change my ways. Even before reporting in, I was of a mindset that I would struggle academically. It was true that I had never really enjoyed being in a classroom, and I was appalled during Swab Summer at how eager many of my classmates were to start the fall semester. After we took the first placement exams, I cemented in my mind the idea that I would be an academic outcast.

In high school, my parents had made it clear that I didn't have to get straight-A grades, but anything less than a B would be cause for concern and likely lead to remedial steps on their part. I feared this veiled threat meant less surfing and socializing on the weekends, so I found a way by which I would go through most of high school with grades on the lower end of the B spectrum. On more than one occasion, my mom would shake her head at my report cards and wonder out loud how I had devised the least amount of effort required to stay above the cutoff for a B becoming a C. I was fully capable of getting an A, but only in classes where I felt the teacher or the subject warranted my full and focused efforts. In short, I was an obstinate, and at times, mischievous youth.

This attitude followed me to the academy. I wrote to my parents that my advisor, my company officer, and my teachers told me that *my major is my 'most important subject,' nautical science is also the 'most important,' macro should be my 'hardest class,' and calculus and physics are worth the most credits.* I then quipped something about how I felt tennis was underrated and should, in my opinion, count for more credits.

A 2.0 GPA became an unspoken goal of mine. Convinced at that point that no amount of effort would do any good to get my grades up, I wrongly focused on the 2.0 and would struggle for the remainder of my time at the academy to do much better than that. In high school, I had prided myself on walking a defiant fine line. Here, however, the risks were far greater. Being told that I couldn't sail again until my grades improved was a shot across the bow that I desperately needed. In the same way that I had resigned myself to lackluster academic performance,

I was determined to exceed any expectations as to what I could bring to the sailing program. Something that I truly wanted, to be a part of the varsity sailing team, had been temporarily taken away from me, and I was determined to get it back. I was also now even more determined not to make that same mistake again.

As October ended, the first snowfall had arrived in New London. I had two trips to look forward to. The first would be over the long weekend to observe Veterans Day. The second was a snowboarding trip with Chuck that we had managed to plan over the previous few weeks. I was, as usual, barely keeping up with my classes and spent much of my free time badgering my teachers for help during their office hours. This continued persistence remained the key element of my poorly planned and more poorly executed strategy to pass classes. I had nicknamed my calculus professor, 'Dr. Derivative' and vented my frustration with him in almost all my correspondence with my parents.

While afforded some new freedoms as a third-class cadet, much of my free time was still devoted to the same non-academic endeavors as my fourth-class year. Uniform preparation always took a fair bit of time. Pants and shirts were ironed with ribbons, nametags, and shoulder boards arranged neatly, and shoes were shined. Our rooms required a fair bit of attention as well. At least monthly there would be a formal room inspection on a Saturday morning.

The preceding Friday nights were spent cleaning every nook and cranny of our spartan rooms. We were also expected to wax and buff our floors, which I found to require some forethought to accomplish on time. For the 100 or so of us that made up a company, there were perhaps two functioning buffers. Moreover, it was not uncommon for the fourth-class cadets to run a late-night mission whereby a broken buffer belonging to our company would be swapped out with a working buffer from another company on the eve of an inspection. There was almost always a mad scramble for buffers in the half hour leading up to an inspection, made even more hectic when it was discovered that one was not working, thereby leaving the company with only one. With each inspection, there were bound to be some rooms with floors unwaxed or unbuffed, an unforced error that would result in some admonishment or demerits. Added to the cleaning routine were my daily runs to keep my sanity, plus a trip every weekend down to the laundry room to prep for the upcoming week. The day-to-day routine had not changed much from the previous year.

In some strange way, I looked forward to any activity that didn't require studying. Because of that, I did notably well on most, if not all things that fell outside of academics. A fair number of cadets would slip out for some liberty on a Friday night, delaying cleaning their rooms until later that night, or the following morning. This usually did not fare well for them during the inspections, but my room always seemed to meet or exceed the current standard, and I avoided the ire of my company officer. In addition to our grades, we were also graded on our military performance. This consisted of a running tally of any missed formations, poor uniform or room appearance during an inspection, or any other times when an officer felt it appropriate to 'award' a cadet with demerits for some error.

I worked hard at presenting a clean uniform, a well-kept room, and made it a daily goal to stay under the radar of the company officers. For my efforts, I kept myself towards the top of my class in terms of military ranking. This, I thought, juxtaposed nicely against my academic standing near the bottom of my class. It wasn't all that difficult to iron a uniform, clean my room, or show up on time when and where I was supposed to be, but I figured that the effort might count for something. I continued to watch cadets get in trouble for sneaking out at night, staying out too late on the weekends, or the inevitable alcohol incidents that seemed to occur nearly each weekend. I may have been delusional by that point, but my belief was that the entire cadet experience was a war of attrition, the end goal of senior officers being the complete destruction of the Corps of Cadets. I wasn't going to let them take me out with something so simple as staying out too late. And with that, I was surviving, which I was convinced was all I could do. It was also all I was concerned with accomplishing in and around Chase Hall. Down at the waterfront was where I hoped to shine.

Veterans Day fell on a weekend that year and gave us a three-day weekend. I managed to secure the entire weekend off and flew down to Texas to visit Leslie at her college. To say I was excited would be an understatement. First, I missed seeing her and knew we would have a good time together. Second, I would be away from the academy. Third, this would be my first experience at what I felt was a "real" college. It may sound ridiculous, but I craved any opportunity to forget, even if for only two days, that I was a cadet at the Coast Guard Academy.

We wandered around San Antonio that first night, found some place to have a long dinner by the River Walk, and had her dorm room to ourselves for the weekend. During the brief periods of time when I wasn't entirely consumed by how much I enjoyed Leslie just being herself, I took in the surroundings of her school and the normal college kids that seemed to carry with them a free-spirited approach to every minute of the day. I tried to put myself in their shoes and wondered how I'd feel sleeping in and casually walking to class mid-morning in a dirty pair of jeans and a t-shirt. It seemed otherworldly.

Experiences like those are double-edged swords. I'd looked forward to that weekend with Leslie for weeks, if not months. But it was just that, one single weekend that came and went faster than either of us would have liked. The truth was she was wrestling with a lot of difficulties as well, hers being far different from mine. Over that weekend, I first felt something standing in between us. It was small and insignificant at first, so small that neither of us found it too difficult to ignore. It wasn't something that either of had done, or not done, but whereas every other time I'd seen her, I felt us growing closer, there was now the first hint of something in the way. We talked for hours each day and night, but seemed to subconsciously dance around whatever was in the way. In hindsight, it's unlikely that anyone's college years don't bring with them some growing pains. Mine, however, were numerous, immediate, and seemingly growing with each passing day.

Leaving at the airport for the flight home was god-awful. Not only because I was saying goodbye to Leslie for a while, but also because I'd blown off studying for the weekend and knew that I would pay my dues the following week. I left San Antonio with a new cowboy hat and a heart that felt like someone was stomping on it with each step I took towards the boarding gate. I was a walking cliché country song both in appearance and mood.

By mid-November, I was complaining to my parents nearly every night about the onslaught of Physics, Calculus, and Macroeconomics. I was trying much harder in Nautical Science by this point, as I'd learned that I could offset my poor grades in math and science with a higher grade in Nautical Science. I took, and failed, another Physics test, reporting to my parents with the upbeat news that while I had failed it, I had failed it by 22 points less than the previous Physics exam that I had failed.

My emails home were generally sarcastic, but not enough to hide the ever-present fear I had of being expelled. In between weekends away and holidays, I fell back on running as a check-valve to manage the stress in my life. At the time, we were issued mesh shorts, cotton t-shirts, sweatpants, and sweatshirts for physical fitness attire. In the thick of a New England winter, that was wholly inappropriate clothing to deal with the puddles of melting snow and ice that covered most of the roads and sidewalks.

Nevertheless, one of my preferred routes was to run out the front gate of the academy, then jog along some of the side roads around Connecticut College. If I was lucky, I could avoid being splashed by passing cars, but often by the time I was back in Chase Hall, my shoes, socks, and shorts were soaked, my sweatshirt not much better, and my skin bright red from the stinging cold as I made my way to the showers. No doubt the passing drivers thought I was out of my mind, and some psychologists would certainly have had the means with which to diagnose me as such. However, I ran with that same stubbornness that would serve me well, both at the academy and in my career to follow. If nothing else, I was well-schooled in being cold and wet.

Chapter 7

Late fall of my third-class year was taking a toll on me. Sailing was over for the season, the spring season too far in the distance for me to even think about, and ocean racing was even further beyond any horizon I could see through the grey New England sky. Calculus had never made much sense to me, ever since I'd seen a letter where a number should have been, and Calculus II was not making any sense at all, no matter what I tried. I spent hours in the library at night trying to rework what I'd just seen the teacher do that day, but it was almost always to no avail. If I was able to follow along in class, it was just barely. Hours later, that evening, when I finally sat down to study, too much had transpired between then and now for me to put the pieces back together.

Physics was no better, nor was my Macroeconomics class. In Nautical Science II, I was applying myself as best I could and hoping for a grade that would offset whatever damage was inflicted by Calculus, Macroeconomics, and Physics. As Christmas Break neared, I managed to pass a Calculus exam and did well on a Nautical Science test covering navigation rules. I had learned quite a bit of Nautical Science just from sailing. Boats, big or small, are obligated to follow a set of international guidelines: larger vessels generally have the right of way over smaller ones, sailboats generally have the right of way over those with engines, and so on. Some of these guidelines could get complicated. There were various configurations of lights for each type of boat and then for the type of work they were doing. We were expected to know all of them. Before graduation, each of us would have to pass a test that was, for many of my classmates, a dreaded event. I, on the other hand, having spent a good deal of time sailing on both open and coastal waters, had learned the basics, and needed little more than to review the more technical sections on ships' lighting to be well prepared. It was far easier for me to learn something out of a book after having seen its practical application and sailing had given me ample opportunities on the water.

As December began, if I could survive the semester, I'd be home for Christmas in just over two weeks and would see Leslie for about a week. The content of my emails home to my parents centered on where my grades stood and making sure Christmas presents that I'd ordered for Leslie had arrived. With some light at the end of the tunnel, I kept up with my daily runs, writing home on the second of December how odd it was that my nose seemed to bleed anytime I ran in temperatures below 30 degrees. That I ran with blood coming out of my nose did little to comfort my parents, but running was a daily ritual for me, something I could look forward to at the end of the day to break up the monotony of classes and studying.

The silver lining to my obsessive running was the ease with which I went into our semiannual physical fitness exam requirement. It consisted of the usual array of sit-ups, pushups, a shuttle run, and a one-and-a-half-mile run. During our first summer, nearly everyone did well on the assessment. But as time went on, and as direct supervision of our fitness waned, many of my classmates opted to forego exercise for sleep or studying. Twice a year, we were herded onto the indoor track where a handful of coaches would have the various stages of the test set up.

Pushups and sit-ups were straightforward events that were usually followed by pullups. I learned early on that the easiest way to up my score was to stop on the lower soccer fields at the pullup bars along our obstacle course and crank out as many as I could while out for a run. I'd usually rest a minute or two then go for another set, followed by a third. During the test, the max score was 20, and the coaches would judge whether each repetition counted as a full pullup or not. On my own, I was able to get to 20, but for the duration of my cadet career, I was never credited with more than 19, owing to the coaches not counting one or two attempts where I may not have lowered myself down with arms fully extended. Each semester, I'd go into the test certain I'd get 20, and each time I walked away frustrated with what I felt was Olympic level bias on the part of the 'judges.'

My weakest event was the shuttle run and standing jump, which were less about distance than they were about explosive speed out of each leg of the test. High-level athletes can be tested to ascertain the amount of fast-twitch versus slow-twitch muscle fibers in their bodies. Fast-twitch muscles help in the anaerobic events like a sprint or jump. Slow-twitch muscles are better suited towards long distance running or bike riding. Lacking any scientific evidence, I decided early on that I was almost entirely comprised of slow-twitch muscles. Once I got up and running, I could get going pretty quick and keep a good pace. But off the starting line, I was always slow. In the shuttle run, my results were horrible. The old adage of 'slow and steady' had no place in the shuttle run, and neither did I.

The muscles in my legs matched the speed at which my brain seemed to process subjects like physics or calculus. I was capable of learning, but not at the speed at which my classmates seemed to pick up a new concept or formula. Thankfully, the last event was the mile-and-a-half run and I went into it knowing I would finish near the front of the pack. Over the nine or so minutes of running, I would work my way slowly but surely up to the front, shadowing someone in front of me for a few seconds before calling up the reserve strength that I had to sprint ahead to the next runner. I could never quite reach the guys who were on the cross-country team, but I wasn't that far behind them either.

Scoring high enough on the fitness test would earn a cadet a blue star to be worn on their uniform for the following semester. My run, pushups, sit-ups, and pullups were all high enough, but the jump and shuttle run always knocked me down a few notches and I never quite got the score high enough to be recognized. At the back end of the spectrum, many of my classmates were subjected to remedial fitness training and testing after having failed the assessment. In keeping with my goal of maintaining a low profile, I walked out of each fitness assessment ranked comfortably in the middle of the pack, knowing that while no officer was going to shake my hand and tell me 'Good job,' none were going to pull me aside for counseling.

In most of my communication with my parents, whether it be emails or phone calls, my mom took the brunt of my complaining. I tried a thin veil of humor in a half-hearted attempt to mask the unhappiness of my days, making comments about the rain, the cold, the wind, or the various nicknames I had come up with for my instructors. At the time, I had little insight into how much of a toll that was taking on them, but my mom took it all with patience and grace.

Her replies were enthusiastic and empathetic. At times, when she was unsure of how to respond to a particularly bad day, my dad would chime in, often from his work email. He was, at that time, nearing 30 years of active-duty service with the Navy. There can be no doubt that he'd seen and experienced things on par, and far worse, than my experiences at the academy. He'd survived the Naval Academy, Naval Flight Training, and a career as a helicopter pilot with tours as the Commanding Officer of several squadrons and Naval Base Guantanamo Bay. I didn't know it at the time, but he'd endured his fair share of setbacks and disappointments along the way.

During his tour as the Commanding Officer of Guantanamo Bay, we had only been there for a few weeks when an uptick in migrant activity in the Caribbean led to tens of thousands of them being housed in camps on the base. When the security situation deteriorated, higher-ups in the military made the decision to send all the American families living on the base back to the United States. Prior to that assignment, my dad had spent significant time away on deployments, and we'd hoped that the two years at Guantanamo Bay would provide a better work-life balance.

This plan was cut short and at age 13, I waved goodbye to my dad and boarded a plane for home, after which he was left to serve out the remainder of his assignment unaccompanied. Over the next year, we would see him once or twice over the holidays. It was a tough assignment for him as well, constant political interference with an ever-changing migrant crisis only made the job more difficult. I never heard him complain once.

Yet here I was, in an entirely academic setting, complaining to my parents on a near-daily basis about how much my life sucked. After one particularly bad phone call, my mom had no doubt called in some fatherly firepower. The next day, my dad emailed me from his office, writing, *"When you go after something as demanding as you have with USCGA, you will find yourself in pressure spots from time to time. All you need to do is take a deep breath and press onward."*

Years later, my dad relayed a story about a friend of his who had flown A-1 Skyraiders during the Vietnam War. His takeaway from combat gave him a unique perspective on life. No matter how bad a day seemed, so long as your engine wasn't on fire and little men in black pajamas weren't shooting at you with rifles, things really weren't all that bad. In my case, no one was chasing me with rifles; rather it felt as if I was constantly running from professors' saber-rattling with rulers held in clenched fists. This was, at least, the perspective I had at the time.

My dad's advice from the beginning had been consistent. Take things one day at a time, and when the day's challenge seems insurmountable, take even smaller steps, and try to simply make it from one meal to the next. Taken as a whole, the four years seemed an impossible challenge. But broken down into smaller chunks, making it to lunch or dinner seemed like a distinct possibility. The same was true with offshore sailing as I often thought no further than an hour or two ahead when I could go down below to thaw out my feet and get some sleep. Unknown to me, I was formulating habits that would serve me well for the rest of my life. Persistence pays off and if you don't allow yourself to think about quitting, you won't.

As December pressed on, I was down to the wire trying to manage a passing grade in Physics I, Macroeconomics, and Calculus II. Add to that my efforts to get as good of a grade as possible in Nautical Science, and my plate was full. Christmas Break was on the horizon, as was spending a week or so with Leslie, but the light seemed at best to be fading in and out.

With less than two weeks remaining of the fall semester, I hunkered down and made one last push to survive. I wrote a lengthy email to my parents about the seemingly massive number of cadets facing disciplinary action for things ranging from cheating to sexual harassment to alcohol. In some of the more salacious incidents, all three elements seemed to form the perfect trifecta for getting oneself kicked out.

One of the many tests we took outside the regular academic schedule was based on Coast Guard regulations. At the time, we were taking the same tests senior enlisted folks took while they were advancing throughout their career. While finding the derivative of the letter 'x' buried deep in some formula was a bit too theoretical for my taste, rote memorization of the uniform regulations or specifications of our ships and aircraft seemed a bit more practical and something I could more easily wrap my head around. Some in the senior class had been caught cheating on their final test. Sexual harassment charges were pending for several of my classmates, and some other cadets senior to us. A fourth-class cadet had accrued multiple alcohol incidents on top of failing grades, and it was a foregone conclusion that he would not be returning after the holidays. Rounding those out was an incident that allegedly involved one roommate being inebriated and urinating on or in the vicinity of his roommate's possessions.

Yet another of my classmates, who at the time had close to a 4.0 GPA, had posted pictures of girls in my class to some start-up website where anonymous visitors could rate them on a scale of one to ten. This of course spread around the academy like wildfire, and he was quickly outed as the perpetrator. It did not help that he chose less-than-flattering pictures for their non-consensual internet profiles. He was expelled. I wrote to my parents about his grand plans to move out west and work as a ski-lift operator. In my mind, I wondered about two things. First, how would it feel to have a stress-free life as a 20-year-old working at a ski lift surrounded by snow-capped mountains? And second, how could he be so stupid to waste the opportunity in front of him? At the time, I would have sold some of my less-important organs on the black market for a 4.0 GPA and here he was, apparently so bored with his free time that he was poking around on the internet looking for ways to have a laugh at someone else's expense.

I didn't revel in others' misfortunes, but I felt that when others had gotten themselves into trouble, that left less time for the officers to pay attention to me and my lackluster performance. I developed a theory—which was most certainly false—that there was a finite amount of trouble that the academy could or would handle at any given time. There was always plenty of it to go around. As I refined my theory, I postulated that once that finite amount of trouble—alcohol, fraternization, sexual harassment, etc.—had occurred, there would be a corresponding reduction in the likelihood of any minor infractions getting an officer's attention. This was complete nonsense as the academy no doubt had plenty left in its reserves to deal with errant cadets, but it made me more confident each day that so long as I didn't partake in any of the shadier activities going on around the campus, I was relatively safe.

Safe, that was, until I had another test to take. I wrote to my parents to report that I'd dutifully failed the last Calculus and Physics exams. *"I always feel a little better after I fail something. It is worse when I have to contemplate failing. The act of failure is not as bad as actually comprehending your faillabilities. I made that word up."*

My goal then was to minimize the bleeding on my final exams and hope that my professors would show mercy on the grading curve. Surely the Coast Guard desperately needed officers, even if their math and science skills weren't quite what their country asked of them. I had some good plans lined up for the winter break, first spending a few days with Leslie before hopping on a flight to Texas to link up with Chuck and his family. From there, we'd drive northwest, eventually making it to Colorado for a few days of snowboarding. I emailed my parents mid-December with a minor hiccup in that my good friend Jon McFerran had taken a fall a few weeks earlier on a ski trip and banged his head. Knowing Jon, he had probably ended up in a hospital. (Over the years, Jon went to the hospital a few times.)

This of course became a big deal at the academy and the higher ups announced a new rule where any cadet skiing or snowboarding was required to wear a helmet. We reckoned that the intent was not to prevent head injuries, but rather to create a rule that would make it easier for them to punish us if we damaged our brains. With only a week or so to go, I had no helmet and no way of acquiring one without the help of my parents. So in addition to reporting on my failure of yet another Physics exam, I kindly requested some assistance in the procurement of a helmet to protect the mostly empty void that was my head.

Perhaps to cheer me up, my dad emailed me with news of a big rescue that had taken place not far from their home. Roughly 200 miles off Cape Charles, an HH-60 Jayhawk from Air Station Elizabeth City hoisted 26 crewmembers off a sinking cruise ship. A second Jayhawk followed not too far behind and picked up the remaining survivors. This was notable as it was believed to be the largest number of people hoisted into an H-60 helicopter. It also caused some international controversy as the sinking was suspicious. Apparently, the ship had been the subject of some shady business dealings and carried with it an insurance policy that greatly surpassed its scrap value. Thankfully, Air Station Elizabeth City had rescued the entire crew. I didn't know it at the time, but I would go on to serve two tours at Air Station Elizabeth City during my aviation career and see my fair share of notable cases.

I rode the train home again, arriving just in time to head to the airport and pick up Leslie. We spent the next few days around Virginia Beach, and it was enough of a break for me to forget New London for a while. We had developed a habit of never really making concrete plans for the day. It didn't seem to matter what we did; we were going to enjoy the time together regardless. We'd been dating for half a year at that point yet each time I saw her, I was hit with the same flurry of emotion that I'd felt when we first met. Her ever-present smile and laugh were contagious and told me she felt the same. There were times, though, when her thoughts drifted back towards the realities of being a freshman in college and all the challenges that came with that. At that point, simply holding her hand and listening was enough to overcome the pressure she was feeling. Whatever it was, we talked around it, neither of us knowing any other way and neither of us wanting to drag the other down.

After Leslie flew back to Texas, I hopped my flight down to Houston and met up with Chuck. From there we drove to Albuquerque, spent a night, then on to Colorado where we managed to have a memorable time snowboarding for the better part of a week, with our helmets on of course.

With the fall semester over, it was easier to forget about academics during the break. For those two weeks, I didn't have studying or tests or labs hanging over my head. On the other hand, I wasn't entirely sure I'd made it through the semester unscathed. As I would find out after returning to the academy in early January, I had passed all my classes, albeit with a D in Calculus II, Physics I, and Macroeconomics. This put my overall GPA into a precarious position. And with another heavy semester in front of me, I was not at all excited to start back up.

On a brighter note, I had made the "Commandant's List" and would have a small silver star over my name tag on my uniform to denote it. There was some formal recognition for what the academy deemed to be notable performance in several categories. The "Commandant's List" recognized those cadets who were ranked in the top percentile for overall military scoring. This essentially meant that I kept my uniform sharp, marched well, kept my room clean, and I was consistently on time and out of trouble. There was also academic recognition that earned those with high GPAs a gold star. For that, I didn't even come close. Then there was a blue star for those cadets that scored in the top percentiles of the fitness evaluation that we completed each semester. My inability to ever score 20 pullups still bugs me to this day and I never did quite score high enough to rate the blue star. Lastly, there was a bronze star that recognized cadets who improved their GPA by .5 or greater from one semester to the next. While I was none too thrilled to be getting a D in anything, I didn't realize at the time that I was setting myself up quite well for future recognition.

Chapter 8

I returned to New London shortly after New Years' Day. Early January in Connecticut is a tough time of year, the days being short and the weather being generally cold, grey, and windy. I was recharged and more ready to tackle another semester than I expected. Leslie's mom had sent me some pictures that she'd taken of Leslie on the beach in Texas and I had them spread out and taped to one small shelf where we were allowed to have personal things out in our room.

Somehow, I had managed to pass Calculus II and celebrated my D by walking down to the river on a Friday afternoon and throwing my textbook into the icy Thames. It sank unceremoniously, but I was overcome with joy to never again wrestle with letters pretending to be numbers. With a new semester not yet begun, I was not bogged down by impending tests and projects. It was a short reprieve from what would inevitably be another tough semester. I was also at this point on academic probation, which just meant I was on a naughty list somewhere and forbidden from going out on Friday nights. The thinking among the faculty was that cadets on academic probation would then be studying on Friday nights rather than socializing. But every Friday evening I was exhausted from the week. I'd order a pizza, eat it, watch whatever movie was on in a friend's room, then go to bed. So much for that additional studying.

January also had a long weekend to observe Martin Luther King Day. Chuck and I went with a friend of ours, Josh, to his parents' house in Connecticut. We had a routine on long weekends during the winter months where the three of us would snowboard at a small mountain near his house each day and sneak beer in his dad's basement at night. Over time, we also worked our way into the social circle of some girls from Josh's hometown. I wrote to my parents in late January that Josh was now acting strange, cleaning the bathrooms and lighting candles before one girl, Kim, would arrive. Chuck had, by this point, developed some annoying traits, also when in the presence of college girls. He was never content with just one. So while Josh set the mood with candles and Chuck ran around like a romantic wrecking ball, I stood on the sidelines with a drink in my hand to watch the show. I was consumed with thoughts of Leslie and called most nights to talk with her before going to bed, so I laid low while the antics played out in Josh's Dad's basement.

Back in New London, the 800-pound gorilla in front of me was Physics II. But the silver lining was three classes, American Foreign Policy, Comparative Politics and Organizational Behavior. By their titles alone, I was certain that none of those classes would present questions with only a single correct answer. I reckoned that America's track record on foreign policy was proof enough of that fact. Class participation meant very little in Calculus, Chemistry, or Physics, but in something like Comparative Politics, if I could present an articulate point from time to time, this act alone would boost my overall grade. These kinds of squishy subjects were abhorred by most of my classmates who gravitated towards their calculators, batteries, and beakers. I, on the other hand, loved the squishy stuff.

To get us through the darkest months of the winter, the academy leadership would occasionally make attendance at sporting events mandatory. This was done in the name of morale and as I would learn throughout my career, 'forced morale' was in fact a thing that rarely nudged the barometer for anyone's quality of life in the intended direction. There were weeknights where I attended several wrestling matches and basketball games, all the while thinking how I'd rather be in my room in my sweatshirt and shorts, studying for whatever test I was soon to fail. I still attribute my overall disinterest in sports to those cold walks in the night air down to the athletic center to watch people bounce a ball around. For a guy on academic probation, there was little satisfaction if 'we' won or lost a game.

By February, I was deep into another semester of physics, writing to my parents, *Saturday night. At my desk. Studying physics.* Making matters worse, I learned a few days later that the show *Baywatch* was being cancelled. I was not entirely kidding when I said that the show played a major part in my interest in the Coast Guard, and now, in the darkest depths of a New England winter, far from Leslie and the warmth of south Texas, weighed down by dull classes, and with grades that reflected my attitude towards them, I was sullen. Sailing, the only thing I really enjoyed, was still too far out on the horizon to even think about.

As our third-class year moved along at a snail's pace, my class was presented with additional etiquette training. During our fourth-class year, this had focused mostly on what fork to use with one's salad at a formal dinner party and the correct method by which we were to dip our spoon in a bowl of soup (away from ourselves in case you're curious). Our main privilege at the end of our fourth-class year was deemed 'Carry-On,' which meant we'd no longer be required to run everywhere and march in the barracks. At the end of our third-class year, so long as we'd collectively behaved well enough, we'd be allowed to keep civilian clothes, like jeans and t-shirts, in our rooms and we would have the privilege to wear them during liberty hours on the weekends.

For much of my fourth-class year, just getting off the academy grounds to do anything was a treat and wearing our uniforms didn't seem that bad. As I crossed the halfway point of my third-class year, I found myself increasingly tired of going out on liberty in uniform. We were also allowed an alternate outfit which consisted of those dreadfully pleated khaki pants, a dress shirt, tie, and sport coat with leather boat shoes. This, to me, looked ridiculous and I refused to wear it. As the spring wore on, I was not leaving the academy much at all as I preferred to simply lounge around Chase Hall on my off-hours in a pair of sweatpants and a sweatshirt.

Nevertheless, our etiquette teacher, who I described to my parents as a *pig-headed cow-woman creature,* got us for an hour early one morning to educate the class about proper civilian attire. Her focus was on the importance of having a nice sport coat and possibly some flannel shirts for more casual and wintry settings. She also recommended pleated slacks. For a kid that had grown up in a wet pair of boardshorts and a tank top, I was appalled at what she was telling us would be appropriate leisure attire. I wrote to my parents that I had secretly ordered a t-shirt from a punk-rock band depicting a man in a diaper proclaiming, 'I don't want to grow up' and how I planned to wear it to her office on the first day we were allowed to have civilian clothes.

In addition to falling increasingly behind in classes, I had to fill in on some weekends with random duties that were assigned to each company. In late February, I was assigned to answer phones outside the admiral's office for a morning. Thankfully, the admiral was not at work that day, apparently having taken an opportunity to do some cross-country skiing. His staff, however, was busy at work. Writing again to my parents, I explained that there was a robust debate about what to serve as the main course at an upcoming reception, with the group of senior officers eventually deciding on chicken. As a cadet, this was fascinating yet incredibly skewed insight into the lives of senior officers. I kept my thoughts to myself and blended in with the wall as best I could.

Transferring phone calls was easy enough and my instructions were to take notes and not forward calls to the 'residence' unless James called. I was confused by this as I had been operating under the assumption for the past year and a half that members of the military had long ago divested themselves of first names. Someone kindly explained to me that if 'James' called looking for 'Doug' that it was one admiral calling another admiral and that they were good friends and only used first names. I was thus relieved to learn that at some point we'd have our first names returned to us. Thankfully James did not call to talk with Doug. To round out my afternoon answering phone calls for the senior officers, I ran into my company officer from my fourth-class year. The Lieutenant Commander was visibly shocked and asked how the hell I had gotten a silver star above my nametag. His guess was as good as mine, but perhaps there were some things about the military that I wasn't all that bad at.

As February ended, there were two things to look forward to. First was a planned Spring Break trip to Texas to see Leslie, but this was dependent on her college basketball team not doing well enough to get into some out of state tournament. As much as I cared about Leslie, I wished nothing but horrible performances from her entire team for the next few weeks. The second thing that had my attention was the initial emails circulating around regarding the summer ocean racing season. I would once again be on *Rampage*, which by some miracle had made it through another year without burning or sinking. The season would be much the same as the previous summer, minus the Newport–Bermuda race. We'd begin preps in May as classes were winding down, the first race taking place on the 23rd, which was a 24-hour sprint around the length of Long Island Sound, out to Block Island, and back. The biggest race would be the Annapolis to Newport regatta, a similar distance, but not as treacherous as the open-ocean crossing to Bermuda. We would finish out the season with a week of races out on Block Island before heading out for some mandated training that would take up the remainder of our summer.

I had made it through the first season with loaned foul-weather gear from the academy. At some point, it had been made of quite nice Gore-Tex material and offered significant protection from the elements. The problem I discovered on chilly evenings sailing off New England was that the waterproof qualities of Gore-Tex fade with time. The jacket and bibs that I'd been issued were neither waterproof nor warm. So as the spring approached, I'd guilted my parents into giving me some money for updated gear. I didn't have quite enough for the newest flashiest jackets, so I settled for a smock and bibs designed for inshore sailing that was made of some material that claimed to be waterproof, yet was not as pricey as Gore-Tex.

From the comfort of one's desk, it's easy to surmise that a 400-dollar jacket and 400-dollar set of bib pants is not really necessary for sailing. In Newport the previous summer, we would all often walk over to the outfitter shop downtown and drool over the most expensive foul-weather gear on the racks, knowing that the cost far exceeded our annual cadet salaries. The staff would tolerate us only for so long before shuffling us out the door lest we scare away their paying customers. However, at two in the morning in a storm off Rhode Island, one learns a valuable lesson in how much it sucks to be cold and wet and that there is no price too high for being warm and dry.

My first summer racing had given me a lifetime's worth of exposure to the elements, but somehow I'd forgot that lesson and not pressed my parents for more money, thus relegating myself to another season of shivering nights hiked out on the rail of *Rampage*. I'd be a little warmer and a little drier, but not like I would have been with Gore-Tex.

Equally as exciting was the news that I had secured a spot on the team that would race in a few regattas at the Naval Academy. That involved a trip or two down to Annapolis over weekends to practice before another trip for the regatta itself. Less than a week before Spring Break, Leslie was ecstatic that her team was not going on the road for whatever tournament they'd been potentially going to. I was beside myself to hop a flight down to Texas and we spent the entirety of that week in our own little world. The weather was great, and the week marked a year since I'd first met her on those same beaches. I was aware of a Physics exam the week I returned to the academy but pushed all those thoughts aside to take advantage of every minute I had with Leslie. We idled away more than one afternoon in the hammock on the porch of her parents' house, and in those moments I could not have been happier or further away from the academy. This was a great decision for my short-term well-being but did not do anything to help my Physics grades, which by that point were getting some attention. Once back from Texas, I had to pull myself from the team travelling down to the Naval Academy and buckle down to get my GPA back in relatively safe territory. The light at the end of the tunnel seemed to be farther off than I had initially thought.

Once the spring sailing season kicked off, I was again doing laps around the Thames on the J/22s. With some time under my belt, I was more comfortable moving around the boat and trimming the jib. For most of March and April, it was not uncommon for the temperature to hover in the 40s or 50s, oftentimes dropping rapidly with the last hour of daylight as we practiced our way around buoys and other boats. I still didn't have a proper pair of warm pants to wear, so most often I just wore cotton shorts, and if the rail had dipped in the water on the previous leg, I'd have frigid drops of water running down my legs as I hiked out on the upwind legs. I'm sure that built character. Or so I told myself.

Removed from the team that would travel to the Naval Academy, the season was uneventful. Practices would run for an hour or two at most, then I'd walk back up the hill to Chase Hall, drop my wet clothes to the floor, and take a hot shower to regain feeling in my extremities. Being cold and wet was expected and bothered me less as I came to accept it as a way of life. The smock that I'd bought was enough to keep my core temperature up, but the fingerless leather sailing gloves did little to protect my hands from the elements. I kept a beanie tucked low on my head for some added warmth and tried to warm myself back up on the downwind legs only to turn back into the wind and feel those northerly gusts fill our sails and freeze my face for the next few tacks upwind. Despite the cold, I loved those practices on the water as they afforded me some time away from my troubles back up the hill.

By April, I was writing daily to my parents about physics, detailing to them how little I cared about *boiling electrons, accelerating them through a potential electric difference*, then shooting whatever cosmic substance we'd created into a *magnetic field created by two large coils*, before finally being deposited into a tube which has been *evacuated and filled with mercury vapor*. Of course, this was done to make something glow and thus solve world hunger. What particular gas or liquid that was, to this day I am not quite sure. And of note, at no point in my 20-year career did I ever call upon those skills at boiling electrons to do my job. As comical as my emails seem now, I was quickly losing whatever toe hold I might have had and my chances of passing Physics II were decreasing by the day.

In what should have been an easier class for me, I was also slipping more than I hoped. Foreign Policy was my comfort zone, but I was pulling mostly low Bs and the occasional C+, which would not have been all that bad had it not been for my strategy of using the easier classes to prop up any classes that required the routine use of a calculator. Another class which should have been a softball was Organizational Behavior. It seemed that we would, as a class, spend entire periods debating whether 'selflessness' or 'humility' were more important qualities. In these kinds of conversations, I had a propensity to shine. The problem was that I had little energy at the time to let my inner philosopher come out.

We were also given a personality test, the Myers Briggs Type Indicator, facilitated by the teacher. That's where I learned what an introvert was, as I had been labeled one. Of the 16 possible personalities, I can vividly recall being told that my score was opposite what the Coast Guard was looking for in its officer corps. It was but one of many instances where I was told that I didn't fit in, and as much as it stung each time my academic advisor, company officer, or random professor reminded me, I became even more hardened against the institution. And at the same time, each of those instances cemented my determination to not quit. I ended up with a C in Organizational Behavior, which could have easily been a B had I taken a few deep breaths and put a bit more effort into the final exam.

As the first week of May closed out, I was waiting nervously for the rest of my grades, knowing that Physics could spell the end of my cadet experience. I didn't have to wait all that long for the news that I had in fact failed Physics II. It came as I was still trying to salvage what I could from the easier classes. Within a day or two of my final grade, I made a desperate plea to the Physics professor and offered to come over to the lab on my free time for any kind of extra work I could do to bump my F up to a D.

At first, he seemed like it might be a possibility and asked what I had in mind. It was at that point that I mentioned something to the effect of helping clean up beakers and petri dishes. He then reminded me that those are more common in a chemistry lab than Physics lab. I knew I had blown it. Leaning back in his chair, he looked at me with the kind of disappointment I'd come to expect from almost everyone in academia and told me that my inability to even note the differences between the two subjects was precisely why I had to take Physics II over again.

With final exams over, I waited impatiently not only for my overall GPA, but also for any word as to whether I was going to be kicked out or not. I knew going into the semester that I was not safe and now with a failure in Physics, my days were consumed with thoughts about how I'd be escorted out to the front gate by security and told never to return.

As I waited, we were busily prepping *Rampage* for the beginning of the season. My first assignment was some project where I was fixing up the companionway hatch which needed a combination of some woodwork and fiberglass repair. At any other time, I would have been thrilled to spend my days outside, on the deck of a boat, sanding away at the hatch and using the same skills that the waterfront enlisted folks had taught me the previous summer. Nearly two weeks went by and as we made final preparations on *Rampage,* I still hadn't heard a thing. This was, as far as my academy experience went, the low point of my four years. What I didn't know at the time was that my cumulative GPA had somehow worked itself out to a 2.00, which had I dropped one more hundredth of a point, I would have been gone. For the semester, I had managed a 1.94. But a 2.00 cumulative, as embarrassing as it may sound, was enough for me to make it through another summer. I would have to repeat Physics the following spring, but my second-class year began the moment the class of 2001 graduated, and I was still in the fight. The third-class year was also the last year of the academy's many required classes, so the fall presented a good opportunity for more classes relating to my major and less dealing with electrons, glowing gasses, and calculators.

Graduation was a sunny day and as the class of 2001 tossed their white combination covers into the air, I was mentally checked out, my thoughts focused entirely on *Rampage* as we would leave that evening for Newport to begin the summer ocean racing season. Open water was the perfect remedy for the past few weeks, and I couldn't get down to the waterfront and away from that dock quick enough.

Chapter 9

With graduation over and *Rampage* motoring south down the Thames on her way to our first regatta of the summer, I felt relieved as we slipped under the Gold Star Memorial Bridge and made our way out into the Long Island Sound. Turning west, we had a few hours of motoring in the evening air until we'd tie up in Stamford and bed her down for the night. Having spent the previous summer on that deck, and a few regattas during the previous fall as well, I felt comfortable with the art of keeping an old sailboat pointed in the right direction and off the bottom of the sea. There was little I could do with respect to the engine, but I knew my way around the sails and rigging. I'd taken almost every bit of hardware on that boat apart, cleaned the innards, replaced ball bearings, greased, and then figured out how to put them back together. I also knew some things were not worth taking apart and I'd no doubt broken parts along the way, but that experience now made me feel like a valuable part of the crew.

Once underway, I'd found that my favorite spot to idle away a few hours at sea was as far aft as I could get, leaning against the stern rails and taking in the sights. With adequate foul weather gear and enough layers of fleece, I managed to stay warm in the cool New England air and when motoring, the boat wasn't heeled over or taking on waves, so it was easier to stay dry. I can't say for sure what we did on our first night away from the academy, but we knew our way around most of the coastal towns and wasted little time celebrating the end of another year.

On my crew was a classmate, Matt, who I'd briefly inter-
acted with prior to the ocean racing season. He and I shared the
same sick depraved sense of humor and found ways to make
light of what would otherwise be miserable circumstances.
That summer, we would spend many tiring nights shivering
while trying to keep *Rampage* pointed in the right direction. We
fast became good friends and took satisfaction in the fact that
both of us were unremarkable performers in the academic
arena. We both wanted to prove ourselves in other ways, most
notably as sailors who could fight above our weight class. Brett,
a year ahead of us, was designated the crew chief. The three of
us all got along quite well. With Chuck on board as a repeat of-
fender, we had a solid crew.

As crew chief, Brett was different than Stan. Stan had a quiet
aura and everyone knew he was running the show although he
rarely, if ever, had to remind any of us of that fact. Brett, on the
other hand, was a likable and very experienced sailor, but didn't
slide into the leadership role as easily as Stan had. I watched
with curiosity and found that this wasn't a problem for him, as
he had a natural ability to listen and absorb feedback quickly. I
wasn't thinking of these things at the time, but to juxtapose Stan
and Brett's style of leadership would have been an interesting
case study for any of the academic types up the hill. Stan always
had a plan long before any of us realized the need for one. Brett,
on the other hand, realized the need for action at about the
same time we did, and he mastered the art of relying on our
combined experience to keep *Rampage* out of trouble. I didn't
realize it at the time, but I was learning from Brett's style just as
much as I had from Stan, who had just graduated and was now
somewhere on the west coast as a junior officer on his first cutter.

Making things even more interesting, our safety officer for the summer, whom I had dubbed *the geek* in correspondence with my parents, knew very little about sailing. This was not a problem for us, but it was a problem for Brett, as it was entirely on him to take whatever plan we had concocted and explain it to *the geek*. At times, *the geek* would have his own plan that made little sense from a sailing point of view and Brett would have to run interference between us, the crew, and *the geek*. I, of course, didn't realize it at the time, but Brett was honing a critical skill that would serve him well during his time on active duty. In any profession, there are those that do and those that don't but still think they know better. As the number two on the boat, part of Brett's job was keeping *the geek* from hurting himself or us. Making it particularly challenging was the need to do so respectfully. Brett did an excellent job.

The Block Island race was roughly 190 miles, running east along the Long Island Sound, around Block Island, then back to Stamford. The previous year had taken us just over 24 hours to complete, and this year would likely be the same. In late May, it would be chilly during the day on the water and after the sun went down, temperatures would be downright cold. Armed with new foul-weather gear, I was in much better shape than the previous summer, foolishly thinking for a brief moment that I'd stay warm and dry throughout the regatta.

Rampage, too, felt like she was in better shape, although that was mostly an illusion resulting from my knowledge of where her leaks likely originated and with that, I carried a false sense of security going into the race. The regatta started early the following afternoon and as darkness spread from the east and the sun set behind us to the west, we were met with a strong fuel smell. Down below, from somewhere not easily accessible—no leak is ever easily accessible—one of our fuel lines was leaking.

Small amounts of water sloshing around the bilges picked up the fuel and with each roll of the boat, the rainbow sheen spread further throughout *Rampage*.

The weather wasn't necessarily good, nor was it all that bad, but the seas combined with the raw smell of fuel gave everyone headaches, and I came the closest I had ever come to being sea-sick. Up forward, where we stored most of the sails, I found a bit of relief from the smell as there were no fuel lines that far forward, and there was a partial bulkhead that seemed to block most of the watery fuel from sloshing any further forward. With a few hours of sleep, I was up early in the morning and back on deck to watch the sunrise. Over the course of the day, we made our turn around Block Island and then successfully made our way back to Stamford.

I was not a huge fan of the race itself as we then had to quickly offload most of our equipment to a waiting van before motoring back the same way we'd come towards New London. There wasn't much time to rest and no opportunity for a night out on the town. I never understood the point of a regatta without a night out to celebrate. For the hour or two we spent dockside, it was comical to see the rainbow sheen of fuel that blazed a well-worn trail from *Rampage* to our van. As we carried fuel-stained sails and bags back and forth, there was no doubt to anyone else at the Stamford Yacht Club that the Coast Guard Academy had the least seaworthy boat in the entire regatta.

At the dock, we found whatever clamp was loose, tightened it, and wiped up as much of the fuel as we could before setting off again for a motoring cruise back east along the Long Island Sound. As disappointed as I was to not have a night of liberty, there was something calming about motoring east under the night sky. The pressure of the regatta was gone, the pace slowed, and we could experience the simple joy of just being on a boat. Two more weekends were filled with regattas at Port Jefferson, NY, and Off Soundings, both of which were day races. The yacht club at Port Jefferson was particularly fond of cadets and made sure to show us a good time. Each week in between races was more of the same, fixing things we'd broken and making sure our inventory of spare parts and sails was kept up to date.

As June got into full swing, we were busy prepping *Rampage* for the transit down to Annapolis. The plan was to motor and sail our way down the coast to the mouth of the Delaware Bay, then motor through the 14-mile Chesapeake and Delaware Canal, before being spit out into the northernmost reaches of the Chesapeake Bay and then further south to Annapolis. It would take us a few days of what we hoped would be easy cruising.

The trip out around Long Island and down the coast was an uneventful one. Reaching the mouth of the Delaware Bay, we motored up with an outgoing tide that slowed us down considerably. Split into two crews, Chuck had found a comfortable spot under the forward hatch that afforded a good breeze as he caught some sleep on a pile of sails. Matt and I were taking turns at the tiller, with one of us navigating up the channel and the other driving. We passed several large container ships and, from a distance, could see a large wake behind each of them. Matt, sensing an opportunity for mischief positioned us closer as the next tanker steamed by and rather expertly pointed the bow of *Rampage* into the four-foot waves that now rolled towards us from the tanker's stern. It was fun at first to feel the bow pitch and we laughed a bit thinking of Chuck being tossed around up forward, likely startled by the bow crashing into each passing wave.

Unexpectedly, one of the waves was larger than the others and the bow dug deep into its steep face, sending a foot of water over the rails, and rushing back the length of *Rampage*. Even on the stern, we were soaked up to our shins as the water rushed over the deck and off the sides. *Rampage* shuttered violently and within seconds our safety officer was yelling at us, which he had every right to do. Matt and I were used to being yelled at by this point, so it wasn't all that alarming. What we didn't expect was the open hatch acting like a funnel that had dumped dozens of gallons of saltwater all over Chuck in his sleeping bag down below. As we were getting the last few choice words from our lieutenant, Chuck popped his head up from the forward hatch and we could see that he was soaked. Anyone else that had been sleeping was up as well and, for the most part, laughed at the whole thing, save for our safety officer who muttered something about not being able to trust us. He was not entirely wrong.

Matt and I then decided that we'd try to behave as best we could for the remainder of the transit. Within hours of the incident with a bow wave, *Rampage* seemed to be struggling against the steady ebbing current as we motored further up the bay into what became the Delaware River. In addition to large tankers, the Delaware River was tricky to navigate due to the presence of small islands, breakwaters, and shoals. Minutes later the problem became worse, and we were no longer able to keep any forward momentum. *Rampage* was now drifting south towards shoal water. Matt and I sprang into action to redeem ourselves from our earlier misdeeds. Running up forward, we opened the locker and went about setting up the anchor and anchor line. If we could set it firm enough at the bottom of the 40-foot-deep channel, we'd be safe against the current and prevent *Rampage* from running aground. I heaved it over the bow, and we let the line run until it went slack, telling us that the anchor had hit the sandy bottom. With *Rampage* pulled by the current, we gave the line a few good tugs before letting the rest of it pay out. Then we waited a few moments before Matt realized that the end of the line wasn't tied off to anything.

With it running fast over the side, we grabbed what little was left and tried between the two of us to hold a 40-foot sailboat against three knots of current while both of us desperately tried to tie it off on a bow cleat. Somehow, over our curse words and clenched fists, we managed to tie her off and the anchor held. Our next step was to get some help, which came in the form of a Coast Guard small boat from the nearest station, which was located a few miles up the river. Boat Stations were, for the most part, run entirely by enlisted folks, with a Chief or above usually running the show. There can be no doubt that the station took great satisfaction in having been called out to rescue a sailboat full of prospective officers in training who had managed to break down halfway up a river on an old boat with no business being on the open water.

Thankfully, the crew towed us to the nearest marina, which provided a great opportunity to acquaint ourselves with Delaware City, Delaware. Within minutes of arriving, we had seen just about everything that Delaware City had to offer, and I knew that this was not the town where I would someday settle down. After a quick and somewhat intimidating walk around the main streets, we were back at the marina where a mechanic named Char was tinkering down below with our engine. I can't remember all the details, but Char didn't have many teeth and seemed to laugh and nod in lieu of using words. What concerned me most was that, after an hour or so, he made some indication that we should start the engine up again. Before we did, Char did not want to be down below when we fired it up. I took this as an ominous sign, as Char seemed to know his way around an engine, so I too elected to not be down below.

Without fanfare, *Rampage*'s engine came to life and the throttle seemed to be engaging the propeller. As it turned out, Char was a damn good mechanic. We were soon underway and back to motoring through the canal towards the Chesapeake Bay. Night fell as we exited the canal and motored south, weaving in between the summer traffic of pleasure craft and container ships. Then we passed Baltimore to our west and the traffic picked up, necessitating near constant course corrections to avoid ships, both large and small. There were perhaps four of us totally awake and on watch, the rest of the crew catching some sleep down below. Our standard operating procedures were to not hit anything and play music from a CD player on deck to help pass the time.

Years later, when stationed aboard a 270-foot cutter, I transited these same waters, in and out of Baltimore for a drydock period. On the bridge, there were at least 20 personnel, with even more stationed on the various decks below the bridge. In and out of any port, we'd set a 'special sea detail' which assigned positions to nearly everyone on the boat, the intent being to best prepare and position the crew for any emergencies that might arise. In and out of Baltimore was a particularly long evolution with narrow winding channels and shoals littering the shallow bay. My lasting memories of that and the many other special sea details during my two-year assignment on that cutter were of the tension on the bridge and the minutiae, or what I considered the absurdity, of the entire ordeal.

From the captain on down, there was back and forth bickering over increasingly insignificant details. These were no doubt the same details that I often overlooked in Nautical Science classes and subsequently led to my poor grades. On the cutter, in addition to a GPS position plotted on a computer screen, we maintained a paper chart with updated positions, just like I'd been taught in Nautical Science. On top of that, we used the radar to plot positions and had additional personnel recording bearings to prominent pieces of land. All this raw data was fed to various other personnel whose responsibility was to plot and keep a running log of each position. If that wasn't enough, we had handheld radios, a phone system, and a backup phone system to talk to anyone and everyone stationed at their appropriate positions around the ship. We were, in my mind, in a wartime posture any time we faced the seemingly insurmountable task of keeping the ship within the confines of a channel.

The captain was almost always on the bridge for a special sea detail and would sit high up on a chair that was reserved for them and them alone. The executive officer would also be there, with a radio in hand and binoculars around his neck. He was the enforcer. The operations officer would also be craning his neck to make sure his underlings were doing their assigned jobs correctly. That left the dozen or so of us junior officers to try our best and look like we knew what we were doing—which oftentimes we didn't.

It was rare for any of us to ever actually drive the ship. Rather, there was a helmsman who operated the wheel, and a separate throttleman to move the two throttles. Whoever had the Conn would bark out commands to which the throttleman would repeat each command and execute it. In the times when warships were entirely dependent on the wind, and each course change required dozens of men hauling on the many lines to control those sails, the various positions on the bridge were no doubt necessary. But as a junior officer in the 21st century, I thought it rather excessive to have so many people bumping into each other and yelling on the bridge.

Back on the deck of *Rampage*, as we motored at about eight knots in the same channel that I would transit again years later on my cutter with 20 people milling about, it was largely up to Matt and I to navigate, steer, and coordinate over the radio with all the passing traffic. We approached the William Preston Lane Junior Memorial Bridge in total darkness, where two channels converged on one another and served as a chokepoint for all ships heading both north towards Baltimore and south, back towards the mouth of the Chesapeake Bay. Matt was at the helm, and I was back and forth between the deck and navigation station down below, taking quick plots and cross-checking my position with the GPS. We were on the west side of the channel and seemed to be doing all right when Matt called down and asked if I could take the tiller for a minute. I assumed he needed a minute to relieve himself or get something to drink.

I hopped up on deck, and swapped positions with him as he flashed a grin at me and went below. I assumed wrong, and as I looked up, I quickly saw the predicament that we were headed towards. In front of us, there was a tug towing a barge, beside him a container ship, and one more large ship behind us that was quickly closing the distance. By this point, Matt was standing halfway down the companionway and laughing, as he found great humor in having tossed me a navigational hand grenade.

There exists a clear hierarchy of who has priority over who and these 'rules of the road' had been drilled into me from the first day I stepped aboard an academy sailboat. This particular instance pitched several of those rules against each other in a very immediate manner. I maneuvered *Rampage* and thought little about it, cursing at Matt under my breath while he laughed and slowly made his way back on deck after the risk of collision had passed. It had not occurred to either of us to wake the safety officer who was sleeping below. Rather, at night, in congested waters, we did what the sailing program had trained us to do: think on our feet, know the rules, and take the appropriate action.

Juxtaposing the two times I transited those waters, once on *Rampage*, and again on my cutter, the two instances were similar in details, but vastly different in their execution. The rigidity of a ship's bridge was the absolute farthest extreme of seamanship, rooted both in tradition and military discipline. Situations on *Rampage*, on the other hand, were handled at the lowest level and seamanship on a 40-foot sailboat was far more art than science. We were, by default, trusted to make decisions that, in the cutter world, would rest on the shoulders of the ship's command, not 20-year-old cadets. In the moment, on the deck of *Rampage*, neither Matt nor I really understood just how much we had learned over the past two years.

We slipped into Santee Basin at the U.S. Naval Academy several hours later and put *Rampage* to bed, all of us finding a quiet corner to get some sleep before the sun came up. The following morning, we set about reprovisioning the boat and, as always, fixing whatever things had broken on the transit. Our engine was questionable at best but had held up for the rest of the transit. My parents drove up for the day and I spent an afternoon with them, walking the docks and admiring the boats that the Naval Academy midshipmen would be sailing in the regatta.

Compared to *Rampage*, not only were the Naval Academy sailboats bigger, but they seemed to have had a bit more put into them in terms of rigging and hardware. We were all aware of Tropical Storm *Allison* that had, earlier in the week, made landfall in Texas and its remnants would cross over the regatta's course over the next 24 hours. Looking at one of the largest boats, I sighed and made some remark to my dad that the cockpit looked like it was high enough and protected to keep the crew relatively dry. I still remember him laughing at that one.

We got a full night of sleep before the race kicked off on the Sixteenth of June 2001. With low grey clouds from Allison swirling above us, but the rain thankfully holding off for the time being, I said goodbye to my parents, and we motored out towards the starting line. Compared to the Bermuda race, I felt that a run down the length of the Chesapeake Bay and then a few days up the Atlantic Coast would be a relatively easy affair. I was entirely wrong.

Chapter 10

Tropical Storm Allison caused extensive damage when it made landfall in Texas the week before. Over the course of several days, she meandered her way up through the southeast where she then strengthened to a sub-tropical storm off the coast of Delaware. The start of the Annapolis-Newport race put us just west of the center of the storm, and thankfully on the preferable side of its counter-clockwise rotation. June 16th was a grey and gloomy day as we set off from Santee Basin for the starting line, my parents watching from the docks as we motored out and raised *Rampage's* sails once again.

In 20-knot winds, the starting line was sporty and the fleet quickly set out on a fast-paced downwind run towards the mouth of the Bay. We had a spinnaker up and were making good speed in the protected waters of the northern Chesapeake Bay. As the sun faded to our west, the winds picked up to nearly 30 knots and, at times, gusted higher than that. While most of us were up on deck tending to sails and lines, the confined waters of the Bay necessitated someone constantly plotting our position on a paper chart at *Rampage's* crude navigation station down below and immediately aft of the companionway. The rolling and pitching motion of a sailboat in a storm stood in stark contrast to the air-conditioned calm of the classrooms where we'd all been taught Nautical Science. Add the dim lighting and damp everything and the task of figuring out where your boat is on a chart becomes much more difficult. Brett, an experienced sailor even before attending the academy, took on navigator duties as night fell and he would call up course recommendations and bearings to buoys for us to keep an eye out for. We were not confined to the main shipping channel, as *Rampage* drew only a little over six feet, but there were shoals and unlit hazards that would keep us on our toes until we cleared the Bay and turned north.

Our safety officer, the same one who'd had to deal with me the previous summer (and had tried unsuccessfully to teach me Chemistry), was on deck throughout the evening and yelled back and forth through a small hatch with Brett over the wind and occasional rain to confirm that what we were seeing matched what Brett was plotting below on the now-soaked paper chart. At one point, we were running through three- to five-foot seas and favoring the eastern bank of the Chesapeake with the spinnaker straining against the increasing winds. Tension had been building for some time as we all quietly sensed that leaving the spinnaker up was an increasingly risky move, especially in the dark. It gave us tremendous speed, as we were well on our way to exiting the Bay in less than 12 hours, but the creaking and strains in the rigging also told us that we were pushing *Rampage*, perhaps a bit more than we should.

The lieutenant called down to Brett to double check that the charted depth ahead of us was in fact deep enough. Brett, who was normally quick to reply, paused. The conversation went something like this:

"Brett, are we good up ahead?"

Brett, after a long pause, less-than-confidently replied, "Yes, Sir, I think so."

"BRETT."

Brett, doubling down, replied, "Yes. Sir. Uh, we're good."

"BRETT!"

"Well, Sir, there's one spot that is six feet."

"DAMMIT, Brett...WHERE?"

Silence ensued for a long 30 seconds, before Brett replied, "We're past it now, Sir." In sailing, as in life, it is sometimes better to be lucky than good. Perhaps our timing was such that we'd crested a wave at the exact moment we crossed the shallowest part of that bar. We would continue to test our luck for the rest of the run down the Chesapeake Bay.

Under full sail, *Rampage* was steady, even in the rolling waves. The mainsail and spinnaker were loaded up enough to hold the hull down through most of the chop, although as the bow dug into larger waves in front of us, the rig would groan, and we'd feel her quickly decelerate before the bow would rise up once again and the sails pulled her back up to speed. We were nearing the Chesapeake Bay-Bridge Tunnel, the same one I'd stared out at the night before making my fateful drive up to Connecticut two summers prior. Once we rounded Cape Charles, we'd drop the spinnaker and make a northerly turn towards Newport. I, for one, was relieved to be almost done with the spinnaker. In fair weather, running with a spinnaker was good fun. But at night, in the tail end of a tropical storm, with fatigue setting in, and the boat clearly pushed to or beyond her practical limits, I was no longer having as much fun.

It was at that moment, or quite near to it, when we 'blew up the chute,' which was a cool way of saying our sail exploded. Thankfully it was just the sail rather than our rigging that let go, and we quickly doused what was left of the spinnaker and composed ourselves. Our speed rapidly dropped off. I looked aft from the bow as our safety officer pressed us to get another spinnaker up and into action. This was an even riskier proposition, but as the good foot soldiers that we were, we didn't ask any questions. Matt and I likely mumbled and laughed to ourselves, betting that we'd soon blow that one up too. Several expletive-laden minutes later, we'd somehow managed to wrestle another spinnaker up from down below, correctly rigged it, and felt *Rampage* surge ahead as it filled with the gusting northwest winds. In aviation, there's a term called *cross-cockpit authority gradient* which is a fancy way of saying when the pilot next to you is of a much higher rank, it naturally becomes a bit more difficult to raise concerns of your own and this tends to have a negative impact on the safe conduct of the flight. On *Rampage*, we were all unwilling to speak up about putting up another spinnaker. Had any of us said something, the following outcome may have been different.

The Chesapeake Bay-Bridge Tunnel was now in sight, only a few miles ahead of us. The fastest boats in the race had exited the Bay hours before us and we found out after the regatta that they'd taken a severe beating in the worst bands of Allison. We were only a few hours behind them but thankfully in a trailing position, and the storm was moving northeast and away from us. Still, the winds exceeded 30 knots, and we were holding on for a wild ride. As the Bay widened, the waves picked up and *Rampage* slammed bow-first into larger and larger seas, the rig still creaking and moaning over the whipping wind.

Whoever was at the helm was having a hard time keeping her on course, as each passing wave threw *Rampage* off by a few degrees. With the seas and the overloaded sails, we were now simply along for the ride. I was trimming the mainsail, stationed just forward of whoever was at the helm, and made sure to keep both my feet firmly planted. On top of that, I was able to manage the mainsheet with one hand, so I took the time to wrap my other arm around a stanchion just forward of where I was sitting. This was a good move on my part, as *Rampage* abruptly broached moments later.

A broach occurs when the sails are loaded up in such a way as to overpower the rudder. I'm sure there's a Physics equation to explain the sudden loss of steering, but in layman's terms, the helmsman loses control of the boat, often by no fault of his own, and the lift created by the sails violently rounds the boat up into the wind before it's knocked over on its side. Any half-way decent boat will normally right itself, given a static and calm environment. Unfortunately for us, things were not static. Everything was extremely kinetic. Too kinetic.

I had dumped the mainsail as soon as I'd felt the bow start to turn, but it had little effect. The quickest way to remedy a broach is to dump the wind from the sails, but in our case, it had come on so quickly that little could be done to change what happened over the next few minutes. The spinnaker was also still partially full of wind, which only kept us pinned down, the mast now almost touching the water. I had enough footing and a firm handhold to keep myself from falling overboard and I paid out as much mainsheet as I could to keep the mainsail luff-ing as the boom slapped against and into the water.

Without a doubt, there was yelling and confusion until someone gave the order to drop the spinnaker. With *Rampage* nearly on her side and drifting fast with the wind and current, it was a challenging evolution for the handful of folks that scur-ried up, hand over hand in the dark, inching their way forward to get the spinnaker down, and then out of the water. I focused on my job, the mainsail, paying little attention to what my team was doing as they all went about their respective jobs. Pinned aft with my own hands full, I knew better than to be a backseat driver as most of the crew wrestled the spinnaker up and out of the water.

Each of us knew what to do, and times like these were not well-suited for lengthy discussions about how best to go about keeping *Rampage* from sinking. In the distance, somewhere in the dark ahead of us, I could hear the gong of a buoy, marking the shipping channel that we were drifting through at quite a clip. From my perch on the high side of *Rampage*, I looked out and could see the lights of houses along Chic's Beach, perhaps no more than five or six miles to our south. In that moment, I was wet, cold, and tired, but couldn't help but laugh a bit, knowing that my mom was no more than a few miles from me and would have been quite upset with the predicament I was currently in.

Matt and I were scouring the dark horizon, trying to figure out the location of the buoy that kept ringing as it pitched back and forth in the seas and wind. We couldn't see it but we could hear it, and for us in the back of the boat, there was little we could do but hope we didn't run into it. In time, the spinnaker was down and packed away below. We were close enough to rounding the southeast corner of Cape Charles to forego trying to put it up again. Here the winds from Allison peaked near 40 knots as the storm accelerated further northeast and away from us. Now past midnight, and reaching with the main and jib, we settled in for the last few hours of darkness, waiting on the eastern horizon to show some faint trace of the morning.

As is often the case with storms, Allison passed and the winds died down to almost nothing. Within a day, we had gone from a tropical storm to calm seas. As the afternoon sun set the following day, we sat motionless off the coast of New Jersey, our sails empty and the sea a sheen of glass around us. A small swell crept in from the east but save for what little drift we picked up from the Gulf Stream, we weren't going anywhere fast. There was little else to do other than take in the changing colors of the sunset and watch as hammerhead sharks swam lazy wide circles around us and several other boats from the regatta that were also sitting idle in the still air.

As night fell, the breeze filled in and the rest of the regatta was uneventful. We finished in Newport, taking a few days to repair whatever we'd broken, to include finding a replacement for our blown spinnaker. We had two more regattas for the season, one in Newport and then a week out on Block Island. Now in my second summer season, I had come to really enjoy the week of racing around Block Island. There were several races each day, usually one day consisting of a course sailing around the island, and we were done each afternoon with hours to spare, giving us time to get into trouble around the small town.

Each afternoon, we picked up a mooring ball for *Rampage* with the other boats from the academy adjacent to us, and our safety officers were always quick to disappear back to their hotels. That left us to fend for ourselves, which we weren't the least bit upset about. We had established valuable contacts in town who were able to acquire certain provisions that made our nights more enjoyable. With overloaded dinghies, we'd row ourselves back and forth from the boat to the shore and took part in random celebrations with whatever crew was willing to welcome aboard a handful of sad looking and damp cadets.

Once the season ended, I had several more weeks of training back at the academy before being cut loose for a few weeks of summer leave. Most of it was inconsequential training that would have little impact on our grades or military rank. So long as I didn't get in any big trouble, I'd wrap up a few more weeks and then head south to see Leslie. I laid low and counted down the days. First up was firearms qualifications at the basement shooting range. It was a blur of activity and all I can remember is being reprimanded for shooting my target in his private parts by the Gunner's Mate who was instructing me. Somehow, I managed to qualify with both a pistol and a rifle.

Following some more instruction on leadership for another week, we were flown down to Mobile, Alabama to the Coast Guard Aviation Training Center for a week of familiarization with Coast Guard Aviation. At the time, I knew that flight training was an option for officers, although not directly out of the academy. Since my dad had been a pilot in the Navy, I was familiar with the process. Some of my classmates were dead set on being pilots. I, on the other hand, was solely concerned with not getting kicked out. The thought of two additional years of intense academic training did not sit well with me. While in Mobile for the week, not contemplating a career as an aviator, I took full advantage of the aviator lifestyle and spent most days playing volleyball on the sand courts and swimming in the outdoor pool, two things that weren't possible in New London. There were rumors of nightly trips out into the town, but I was exhausted and homesick, so I made the prudent decision to forego any more adventures and get a good night of sleep each night.

I can't say for certain that a week of immersion in Coast Guard aviation did anything to shape the career track I would choose, but it did leave quite an impression. Someone was no doubt in charge of the two dozen cadets roaming around the Aviation Training Center, but for the most part they left us alone and I was incredibly thankful for that. We had blue flight suits that were leftovers from the gear-issue shop. Most of the pilots and aircrew had all transitioned over to green flight suits like the rest of the military. We stood out as easy targets, but the aviation community seemed cordial nonetheless.

During that week, we spent a day driving over to Pensacola, Florida, where most of the Navy and Marine Corps flight training took place. On the main base, we spent an afternoon roaming around the National Museum of Naval Aviation. Matt was with me, and I took a detour out onto the uncontrolled portion of the flight line where there was row after row of old tired aircraft waiting for their turn at restoration. I knew that my dad had flown an H-2 Seasprite down to the museum at some point in his career as the Navy fleet transitioned to the newer H-60 Seahawks. I found at least one, its paint bleached by the Florida Sun and one, if not all, of its tires flat.

Even in rough shape, I admired the old helicopter and wondered to myself if my dad had flown that particular bird. His time in the Navy mostly took place during the Cold War, when the global order was vastly different than it was for me in the summer of 2001. While capable of many missions, the H-2 was primarily intended to track submarines. For decades the U.S. and Russian Navies went toe to toe around the world, spying on one another and trying to outsmart their opponent. And while the nuclear Armageddon thankfully never materialized, both sides trained for the threat and understood the severity of the consequences if things ever went south, and they were caught unprepared.

By joining the Coast Guard, I had intended to do something slightly different from my father's footsteps. I wanted to be in the military, but in a subtle nod to my dogged individualism, I wanted it to be on my own terms. Standing on the ramp and seeing all the aircraft, I couldn't help but wonder to myself about the odds of successfully completing flight training and being a pilot, like my dad had done before me. Back in Connecticut, my grades were barely enough to even hold onto my cadet status, so I didn't spend too long thinking about seeking two more years of academic pain in the cradle of Naval Aviation.

After an uneventful week, I caught a flight back to Norfolk and had a few weeks at home to recharge and ready myself for another academic year. Leslie and I were together for most of it, and I rented a place down in Cape Hatteras, at the southern end of the Outer Banks. At the time, there was a small Coast Guard unit there, called a Group, that oversaw the handful of small boat stations up and down the Outer Banks. On the base, there were several rental apartments that went for a fraction of what a hotel room would cost in the high summer months. The two of us hung out on the beach for a few days and I did some surfing.

A day or two before I headed back, I drove Leslie to the airport and again felt what had become an all-too familiar awful feeling each time I said goodbye to her. Whether it was for a weekend or for a week, it was never any easier to say goodbye. I was wholly in love with her. The problem, which I had sensed for some time but tried to ignore, was not that she didn't love me, because I was certain she did. Rather, she had a strong desire to break out of the mold she'd been living in for most of her life. She was athletic, wickedly intelligent, and driven, having graduated from high school at the top of her class. She was, at least on paper, tracking towards medical school while playing college basketball at the same time.

Few outside her immediate circle knew she was also battling the pressure that comes along with being an overachiever. No one had ever dared to call me an overachiever, so it was hard for me to relate, but Leslie would offer subtle hints and momentary glimpses into her world. When we were apart, we talked almost daily, often for an hour or so, and as the weeks and months went on, her frustration with it all only grew.

Most conversations, whether in person or on the phone, went something like this:

I'd ask, "What's wrong?" to which she'd reply, "Nothing." Maybe I'd ask a second time with the same answer, and we'd find something else to talk about. Oftentimes whatever subsequent topic we found gave her a means to offer little hints about what she was dealing with. She'd talk about walking to class and include some insignificant detail that offered clues as to how she really felt. It was a frustrating and slow process, made even more so by the fact that there was little I could do to help other than listen. Increasingly, it seemed like listening was not going to be enough.

She had arranged to take a semester of classes in Hawaii at a school that offered exchange programs with her school in San Antonio. I wasn't entirely thrilled with a few more thousand miles between us, but her mind was made up and saying good-bye in August was even tougher for me because of it. In hindsight, I likely knew our time was limited, but at the age of 20 and experiencing love for the first time, I was ill-equipped to handle the burden that came with that kind of emotional complexity. I wanted her to be happy, but more importantly, and somewhat selfishly, I wanted to be part of her happiness.

With my return to New London imminent and Leslie headed back to Texas then Hawaii, I faced another semester ahead of me. I made the trip back up in August 2001, north along the east coast, then through the maze of ramps and over-passes of New York City, before turning east along I-95, into the state of Connecticut, and counted down the exit signs over the last two hours or so before I'd pass through the academy gates once again.

In my third-class year, I'd begun to take a few classes in my major, Government. They'd been mostly introductory, but I'd enjoyed them much more than Physics, Calculus, Macroeconomics, or Chemistry. Now a second-class cadet, I'd could dive a little further into the world of politics. I thought, as most 20-year-olds do, that I had a good understanding of how the world worked. The most recent war that I could remember had been the first Gulf War, which had been a straightforward example of good versus evil. Evil invaded a country, Good went to save it. It didn't take long, there weren't too many casualties (at least on the good side), and victory seemed well-defined and fully achieved. The men and women who'd gone overseas came home to parades and the free world went about its business. This of course all would change not more than two weeks after the academic year began.

Chapter 11

I had just gotten out of a class in Satterlee Hall on the morning of September 11, 2001 when I passed Matt headed in the opposite direction. He mentioned something about a plane crashing into the World Trade Center as we passed and I nodded, thinking to myself that I'd stop by a dayroom and check the cable news before swapping my books and heading back out. The second plane hit just as I was turning a corner towards my company's dayroom, and I could see a large group huddled near the television. I still didn't grasp the magnitude of what was happening.

Classes continued uninterrupted that day, perhaps there were a few announcements made at lunch, but things were relatively normal for us. I was taking a class within my major that was taught by Lieutenant Commander Joseph Vorbach, a teacher I had quickly come to admire. He was intelligent, articulate, and seemed solely concerned with teaching us as much as he could in the short amount of time we had with him. He never came across as trying to employ the common leadership theatrics to impress any of us. This, I concluded, made him someone worth paying attention to during his lectures.

As I sat down that afternoon, it was strange that he was a bit late to enter the room. He was always on time. Minutes passed before he finally entered the room and set his things down. Lieutenant Commander Vorbach was clearly rattled. He understood the magnitude of what was going on, not just in New York City, but also around the world. His facial expressions were different, the sadness of that day hanging over him, and he took some time to choose his words carefully. Whatever subject the class had been scheduled for that semester, we were told that we would instead dive deep into the attacks of September 11th and the impact they would have on the United States and, in turn, the rest of the world.

I leaned forward and strained to hear each word he spoke over the course of that semester and, from that point forward, I signed up for every class of his that I could until I graduated. It was from him that I learned, long before many others did, of the differences between the words *Islamist* and *Islamic*. He was a soft-spoken man who, over the course of a week or so, provided an entirely new reading list for the semester, one that I read cover to cover as I tried to wrap my head around the world that I was now living in.

The entire Corps of Cadets gathered that evening in the Quad and a few announcements were made about changes to security around the campus and what little was known about what was happening two hours away in New York City. Within the first few days, signup sheets were posted around Chase Hall for volunteers to go to New York and help. I signed up with no hesitation, as did most of the Corps of Cadets. Everyone wanted to do something yet none of us had any idea what that was.

Shortly thereafter, we were told to focus on our studies and that our time to make a difference would come later. In our young minds, we felt that whatever was going to happen would be long over by the time we graduated. I was incredibly frustrated to think that focusing on passing Physics II the second time around was somehow going to help solve the more pressing issues at hand. Initially not allowed off-base in the days after, I ran repeated laps around the lower track and stopped for pull-ups, each time finding myself increasingly frustrated that I wasn't doing anything to help.

The wars that resulted from the attacks of September 11th would shape the world in which I would serve for 20 years, if not more. The Coast Guard would adapt to the changing security requirements, both at home and abroad. It was folks like Commander Vorbach that understood that long road ahead. As the days, weeks, and months passed, he taught us about asymmetric warfare, the lessons learned from past invasions of other countries, and the unique realities of fighting non-state actors. Extremism and terrorism, as we came to know both, had no borders. In those first few months, many painted it as a black and white issue, a case of good versus evil. Country Music stars were quick to write sappy patriotic songs espousing the American virtues which would, without even the shadow of a doubt, crush those who tried to destroy us. Luckily for me, I had folks like Commander Vorbach to dissect the true complexity of what we faced, and just how long of a battle it would turn out to be.

As the fall semester of my second-class year began, my father had initially planned on retiring. The events of September 11th changed that, and he extended his time on active duty in the Navy for close to a year. He took on a job in Homeland Security Maritime Planning and Operations for the Navy, not knowing at the time that "Homeland Security" would soon become its own Department within the federal government. In one of his first emails to me after 9/11, he explained that he was often at work while the parking lot was still empty in the mornings and walked to his car early in the evenings when the parking lot was empty once again. He also advised me not to confuse his *.mil* email address with his *smil.mil* one. At that point in my career, I knew little about the world of classified material, other than most of whatever my dad was involved in after the 9/11 attacks was not for public consumption.

My dad would, from time to time, send me links to an article or two to read. Knowing that he was very much involved in the preparatory work and analysis that was going on at a frenzied pace throughout the U.S. Government, I made sure to read and digest as best I could the points being made in each. I wrote to him in early October, *I get the feeling this hasn't even started yet,* to which my dad replied that his biggest fear was follow-on attacks and the impact that would have on the country. First, he pointed out the uncertainty of whether the country would be in it for the long haul, and second, he mentioned the fear that might engulf the entire country. Both were salient points, and I relied on my dad's advice to form my own opinions as time went on.

Within the confines of the academy, I was busy with academics, namely Criminal Justice, International Relations, Naval Architecture, and Nautical Science. As a second-class cadet, I was also supposed to be serving in the capacity of cadre for the fourth-class cadets. Many of my classmates had cut their teeth over the summer as cadre for the Swabs. In fact, most of my class had acted as cadre in some form, either in Chase Hall, at the Waterfront, or in several other programs during the summer. The handful of us from the ocean racing team were the only ones in our class who hadn't spent time marching and yelling at the new class over the summer. I had certainly yelled, but by this point in my cadet career, I reserved raising my voice for things like *Rampage* taking on water or broaching at night in the back end of a tropical storm. Moreover, I'd been yelled at plenty and as I developed my own style of leadership, I saw little value in tormenting someone two years younger than me in the name of tradition.

The sailing season kicked off shortly after our return and I was now trimming the mainsail, once again on *Rampage.* With two years under my belt, and a handful of younger cadets underneath me, that tired old 41-foot hull became my default leadership laboratory. I was too frustrated with Chase Hall to play what I felt were mere games. The cadre would bark out orders and the fourth-class would run around seeing to each task as best they could. Voices would elevate on both sides for no apparent reason and I found that I was happiest when I could tune out most of it.

I was beginning to enjoy some of my classes, but still carried a healthy amount of skepticism for the academic system that had been tormenting me for the past two years. Making matters worse, I had an academic advisor who was trying to convince me to quit. I took a little bit of comfort in rejecting the system as a whole. I could run and salute with the best of them, but where I found the most opportunities to learn and grow as a person and perhaps even as a leader was down at the waterfront. Specifically, I took a lot of pride in the fact the *Rampage* hadn't sunk. I was proud of this fact because we all knew she was quite capable of sinking at a moment's notice, and we'd been at sea enough to give her a fair number of chances to check out the bottom of the ocean. That she had not found the ocean floor was entirely a result of the seagoing prowess we had all developed over time.

With the fall sailing season underway, I began in earnest to explore the role of a leader, making innumerable mistakes along the way. On *Rampage*, I was now in a position where I had a fair bit of experience. I could talk to the junior cadets through tacking and jibing, the common mistakes of each, and the unique challenges of a finicky sailboat well past her prime. *Rampage* was a hot boat in the early 80s, but 20-some years later, advances in sailboat design had relegated her hull and sail plan to the past. I also knew the river, the shallow areas, the way the wind would rip down from the north, and how the currents misbehaved. It was out on the Thames River where I began to build myself back up. I had confidence not because of my grades or academic standing. I was far from the top-performing student in Nautical Science, and surely the butt of many jokes amongst the Physics department staff. But I knew my way around a sailboat, I could grit my teeth through a cold night on deck, and I could take a wave to the face with the best of them.

I'd been to sea, and not in the *Eagle* sense of the term either. Nothing about the past two years of sailing had been scripted or overseen by a gaggle of officers. We'd sailed through heavy weather, fixing broken rigging and fittings with our own bare hands. We'd kept *Rampage* afloat for days on the open ocean when she was trying her best to give up. There was obviously some oversight for what we were doing, but I had a good sense of appreciation for how minimal that was when compared to anything going on up the hill.

Perhaps most importantly, I'd learned a great deal about managing interpersonal relationships. In each regatta, there were no more than eight or nine of us to sail the boat. And each person had strengths and weaknesses; some more than others. It was easy to chide someone for messing up or offering a so-lution that clearly wasn't going to work, and this happened of-ten. But being on a boat, under sail, and in the thick of a race exacerbates the quickness with which teamwork can fall apart. My sense was that the most constructive way to handle that kind of environment was to assume until proven wrong that folks would want to do their jobs correctly. And when they didn't, catching the mistake early and offering a subtle hint to get them going back in the right direction was far more useful than the heavy-handed yelling and shouting in Chase Hall. Stan Hudson had done just this, keeping us all pointed in the right direction, but always allowing us enough room to make and learn from non-critical mistakes. With the challenges we faced and overcame, camaraderie soon followed, even if we weren't always the best of friends off the boat. And that was a contagious phenomenon that seemed to pay huge dividends.

The fall was no different than the previous two years' seasons. In addition to *Rampage*, I got to spend some time on *Steamboat*, a slightly younger J/35. At 35 feet, she handled differently and had some minor rigging differences, but otherwise sailed the same as any sloop would. Roughly halfway through the season, we were in the middle of a regatta, at the southwest corner of Fisher's Island in some fairly stiff wind with the spinnaker flying. As I'd learned, sailing directly downwind poses some risk, and a jibe with the chute up was far more complicated than with a jib alone. At some point, we were downwind and debating whether to jibe as the line we wanted to hold was precarious with the main fully pressed out against the spreaders and the spinnaker straining with the gusting wind.

Unexpectedly, *Steamboat* took a wave that rolled the boat a bit and drove the nose down into the back of a wave in front of us. Almost immediately, the mast snapped just above where the spinnaker pole was attached to it and the rig came down, a jumbled mess of sails, sheets, shrouds, and spreaders coming crashing down on top of us. Half of the mast was now in the water, *Steamboat* was rolling in the passing waves, and Jon Mcferran had been knocked overboard along with our coach.

These were the kinds of things that we couldn't train for but had planned for and talked through countless times. First and foremost was the task of getting both Jon and the coach back onboard, so the countless man-overboard drills that we'd run in the river kicked everyone into motion. As we'd lost our ability to sail, someone already had the engine fired up as Jon and the coach were pulled up and back aboard. At the same time, I hustled down below to grab the bolt cutters that we kept taped around the base of the mast.

With the mast cleanly sheered and now in the water, there was the problem of the forestay, backstay, and shrouds still holding onto it. The sails were also still attached and now underwater, acting as a drogue of sorts. With *Steamboat* tangled in a web of her own rigging, the biggest and most immediate risk was one of the spreaders of the broken end of the mast punching a hole in the hull. The point of the bolt cutters, as I had been told, was to cut the rigging away and let the entirety of the rig sink to protect the hull and keep *Steamboat* afloat. *Fun*, I thought, as I scrambled around the hull, cutting away at the cables to free the rig. On the stern, we had a small hydraulic pump to adjust backstay tension that was about a foot underwater and being pulled by the weight of the rig. I didn't bother trying to save the pump and cut the backstay away inches from where it was bolted into the hull. In quick time, the rig, the lines, and the sails disappeared beneath us into the Sound, and we were left to motor our way home.

Jon had a small slice taken out of his head above one of his eyes, so the coaches made a call to the small boat station in New London, and we teased him as he was soon "medevac'd" off *Steamboat* to get his laceration tended—you'll recall that I said Jon went to emergency rooms with some frequency. The rest of the transit back to New London was uneventful, and we had the added benefit of not having to wait for the railroad bridge to rise up since we no longer had much more than a stub for a mast. Once back at the academy, it was lunchtime so most of us made our way to the wardroom, or cafeteria, to get some hot food. We were surprised to hear announcements from the head table to 'keep 1/C McFerran' in our thoughts as he'd apparently been seriously injured while sailing. Knowing the true state that Jon was in, we laughed and dug into lunch.

Our most crowning achievement during the fall of 2001 was winning the MacMillan Cup at the Naval Academy. I trimmed main for the regatta and viewed the entire event as the most significant race for us, not just for the fact that it was the annual event that pitched collegiate teams against one another, but also in the sense that it was on the Navy 44s, boats that we were not entirely accustomed to sailing. With a classic sloop rig, many of the skills we'd fine-tuned on *Rampage* and *Steamboat* translated well, but the rigging was different, and fittings were in different places with sheets and halyards of different colors. We knew how to sail, but many of the more nuanced aspects of sailing a boat well had to be learned quickly.

Moreover, our crew was mixed from several of the offshore boats at the academy. We hadn't all worked together on the same deck for the duration of the season. With one day to practice, we had little more than a few hours to figure out those critical strengths and weaknesses. What worked for us was the idea of being a team, each of us contributing what we could and accepting constructive criticism of the things we each needed to improve. If the folks up on the bow couldn't drop the pole and jibe fast enough, it was the job of the helmsman and me on the main to slow things down to a pace more suited to the ones doing the grunt work up forward. The other option, yelling at them to move faster, would serve no purpose whatsoever.

We saw this time and again among the other crews, each of them in the same pressure cooker that we were in, trying to figure out a new boat in a matter of hours, then take it to the starting line and race against their skilled opponents. The starting line itself was a grinder, with each boat closing in with the pack as helmsmen jockeyed for position, trying to stay in tune with their crews and push for better and better performance. Frustration, when it inevitably reared its ugly head, could be dealt with constructively, or it would boil over into a counterproductive mess, often taking a team out of contention during a rounding of a buoy or run downwind. Going fast was important, as was correctly reading the wind, but a crew falling apart was the real disaster everyone feared.

That we won spoke volumes about the training I'd received, not just from the academy sailing program, but also from folks like Stan, Brett, and the others who'd taught all of us in our formative years. To be certain, there was yelling at times to get someone's attention above the noise and at other times to solidify the cohesion with which we needed to approach each sail change or tack around a buoy. But still we worked as a team to reach a common and shared goal, one that I was immensely proud of when the final race was called, and our coach gave us a thumb's up from the committee boat that we'd be taking the McMillan Cup home with us.

In December, we took the Deck Watch Officer's exam, with a passing score being a prerequisite to graduate. It was also a prerequisite to be qualified once we reported aboard a cutter. Here again, my time sailing paid off as I managed a grade of 92 on my first attempt. I wrote to my

Winning the McMillan Cup, Fall 2002. As team captain my firstie year, I'd used everything I'd learned from sailing. A crowning achievement.

parents about how, after the exam, cadets paced back and forth in the hallways outside as they waited for their tests to be graded. A handful from my class would have to take it multiple times and a smaller group of first-class cadets were still trying to pass it, having unsuccessfully taken it several times the year before.

I ended up with a 2.80 that semester, which brought my cumulative GPA up to a whopping 2.14, in my mind far above the 2.00 threshold that meant expulsion. It was, all things considered, a very good semester for me. I hadn't seen Leslie for quite some time, so we made up for it over the winter holidays, where I was free for about two weeks before heading back for the spring semester. Sadly, as I'd feared, there was no escaping the nagging feeling that things weren't going as I'd hoped for the two of us. We had a great time together, but she was clearly headed down a road that she needed to take and tagging along for the ride was likely not going to be an option. Looking back, we both probably tried to pretend as if that wasn't the case, but each of us knew something was amiss. With the rigors of the academy, I didn't have a great deal of time to process any of that, my mind convinced that the only way forward was to make things work. Perhaps Leslie thought the same as the two of us began stumbling down the road of a long and painful breakup.

Chapter 12

By early January of 2002, I was back in classes and the spring semester was in motion. On the 16th, I wrote to my parents complaining (as I so often did) about being forced to attend yet another lecture about leadership, this time from the Commandant of the Coast Guard himself. In my words, I described the idea as being *worn and beaten the ground* and that the academy was a *magnet attracting all sorts of people who think they can teach someone else a thing or two about leadership.*

My frustration was born mostly from the constant barrage of the word *leadership* in our day-to-day lives. Any and every opportunity that an officer had to remind us of the importance of leadership was never to be wasted. Whether it was taking out the trash or being early for a formation, anything and everything revolved around being good leaders. I dreaded each of the hour-long lectures in Leamy Auditorium, as it was one less hour that I had to try and figure out how to do math with letters or count electrons. With academics always on my mind during the week, I just wanted to be left alone to study.

More importantly, I was certain that the most valuable lessons in leadership that I'd received had not been in an academic environment. Rather, they had occurred out on the river, or the sound, or the Chesapeake Bay, or best of all, out on the Atlantic Ocean. When we ran into a squall, a line parted, a fitting failed, or we blew out a sail, we didn't all sit down and discuss the intricacies of how we were going to fix the problem. Rather, someone took charge and shot from the hip to get us back on track. In the rain and wind and darkness, there was always a clear understanding of who was supposed to do what. That clarity was born almost exclusively from our endless practices out on the river and experiences we'd had during races or transits. We had a very basic understanding of leadership in that one person could call the shots and it was up to the rest of us to make it happen. This was a meritocracy—if we had an officer with us who wasn't the sharpest sailor, no one needed to say much. We knew that either Stan or Brett had the big picture, and we followed their guidance, leaving the senior cadet to then balance rank delicately with seamanship—or a lack thereof—and somehow keep the peace.

At sea, with a crew of seven or eight cadets, whoever was in charge always ran a real risk of losing our trust. If one evolution at night didn't go well, it could begin to erode the positional authority held by that unfortunate cadet. It's likely that the coaches understood this potential problem and chose the team captains and crew chiefs carefully. Stan had not once lost our confidence. He was a natural leader on and off the water. During my second-class summer, Brett had been the crew chief, and it was a similar experience. We all trusted his nautical intuition and the skillful way he managed whichever officer was onboard with us.

I learned a critical lesson from him. When Brett was not entirely sure of the best course of action during a race, he would consult with Matt for tactics. And when he wasn't entirely sure what we could do with the boat, he would seek out some input from the rest of us. He was subtle about it, yet consistent, and it certainly left an indelible mark on me as I began to feel out my own thoughts about leadership. Asking for others' opinions was not a sign of weakness, it was a smart way to get a diverse mix of answers. Brett no doubt also knew what our individual strengths were. He never asked me about tactics because we both knew I didn't have much to offer. But when it came to a sail change or a tweak in how we trimmed the sails, I was able to offer up some advice and Brett always took it. He was also able to make decisions at critical points in time based on the information that he had, which was most often incomplete. The immediacy of rounding a buoy necessitated quick decision-making, and Brett was always on top of it.

Classes were, for the most part, focused now on my major, but I was also taking a second shot at passing Physics II, which hung heavy on my brain. Failing a second time was not an option and a surefire path to expulsion. One of the many things that made Physics so difficult for me was the required foundation in calculus to run most, if not all, of the equations. I had barely squeaked by in calculus, and now a year after failing Physics II the first time, I was struggling to remember what a derivative was. Dreading each walk to Smith Hall, I dragged myself up the steps to the doors perched high above the river on the eastern side of the campus. I would routinely wait until the last second to enter the classroom, taking as much time as I could in the hallway outside, looking out the windows at the Thames River below, thinking only about sailing. In the dead of winter, the days were mostly grey, with an ominous fog hanging low at times over the water. Despite the brutal weather outside, I could appreciate the raw beauty of it and forget for a moment about sitting through another Physics class.

With the previous semester a win for me, I was able to get off the academic probation I'd been on for the previous semester, which meant I could go out on Friday nights. In the back of my mind, I feared that my newfound freedom would be short-lived, but I did my best to enjoy it while I could. As the winter pressed on, I continued with my normal routine of runs after class, normally taking a circuitous route around Connecticut College to pretend for a few minutes that I was living the idyllic, expensive, and cliché private New England college lifestyle. As I ran, I imagined myself carelessly tossing a frisbee around a manicured lawn with my long-haired hippie friends as we debated the true meaning of Robert Frost's poetry.

I also managed a quick trip down to Texas to see Leslie during a holiday weekend and wrote to my dad about how close I was to quitting and waiting tables in Austin for the rest of my life. Looking back, I think I would have been just as happy in Fargo, North Dakota so long as the pressure I felt from the academy was no longer weighing me down. Just to feel normal for a weekend and walk casually around a town was a relief. Having a girl I cared deeply about and who seemed to feel the same for me by my side was icing on the cake. We were happy in a lot of ways, but there remained some unspoken and ever-present tension. We talked around it, barely skimming the surface of the more complicated things that Leslie was going through. On the surface, she was restless but there was more to it, and while I could see how it manifested itself, I never quite found the right things to say or do to make any headway in offering help. I was far from capable of even knowing what questions to ask or things to say to break through the wall that was now being built between us. I left with plans to see her again in a few weeks for Spring Break, but each day that passed only seemed to reaffirm the sinking feeling that we were not headed in a good direction.

A few weeks later back at the academy, Matt Newell had worked some sort of scheme where he and I flew down to St. Petersburg, Florida to race in a regatta in Tampa Bay. The sailing was fun, but the real treat was getting away from New London in February for a weekend. Matt was an expert when it came to sailing and tactics. I was merely along for the ride and there to trim sails. I'd keep the boat going fast and leave the tactics to Matt. I still often felt like I wasn't as good as I should have been, but in being selected to go on the trip, I was reassured that the coaches seemed to think I knew what I was doing. In addition to two good days on the water of Tampa Bay, I was out running each morning in the 60-degree air, a far cry from the freezing temperatures, slush, snow, and sleet often encountered during my runs around New London.

The biggest news was that I'd been selected as a crew chief for the upcoming summer ocean racing season. There would be two boats, and I'd be in charge—or at least as in charge as a cadet could be—of one of them. Early indications were that it would be *Rampage*, which made sense to me as I was more than familiar with the many ways in which she might spoil a race.

Rampage under power with no sails. I'm a 2/c cadet with Matt steaming home after a regatta, summer of 2001.

At mid-semester, I was holding a 2.50 GPA and seemed to be keeping my head above water. I had passed most of the Physics exams but was struggling in one of my government classes. I was prepared for a D in Physics, but only on the assumption that I would do relatively well in my other classes. The pressure continued as March came around. If there was any relief, it was that the winter was slowly letting go and spring was working to get a footing in southeastern Connecticut. My daily runs went from frigid to just cold as the snow started its annual thaw. Unfortunately, the process wasn't quick as it would melt then freeze again nightly, turning the sidewalks and stairways into frozen obstacle courses. Shin-high slippery stumps of muddied and salted ice littered the campus. There is a reason why scenic paintings of New England are never painted in the early spring.

I also got the news that I would be captain of the offshore sailing team in the fall. That I was a crew chief for ocean racing seemed enough of a reward for the work I'd put in over the years. I was certainly not expecting to captain the sailing team. There were a handful of classmates of mine that were also on the team, most of them in better academic standing than me. Hardly a day went by that I didn't consider my impending expulsion, but the vote of confidence from my coaches helped to put my mind at ease.

I had learned the most from the waterfront, beginning with those first summer days of figuring out how to sail a dinghy, from the first fall season learning how to trim a jib and work as a crew, the endless practices out on the river in the frigid early spring working a J/22 around the buoys with frozen fingers, the practical lessons I'd learned from the enlisted crew in fixing a boat, and the dozens of trips we'd made under the I-95 bridge and out into the Long Island Sound and beyond. To now have such a significant accomplishment ahead of me felt rewarding. I would soon learn that with titles comes work.

With practices out on the river during all but the worst of weather, and the summer ocean racing teams shaping up, there was again that same faint light at the end of the tunnel. But the day-to-day grind of Chase Hall kept me firmly grounded in the present. Even now that I was allowed to wear civilian clothes out on the weekends, I still had a deep dislike of the *etiquette nazi pig cow woman*, as I described her to my parents. At some point, after the third-class cadets had been given the privilege of wearing civilian clothes as well, we (the class above them), were sat down and told to set a better example for them. According to the Commandant of Cadets, who spoke first before her majesty the etiquette queen, we needed to *dress more like officers*. She then took the microphone and added that not enough of us were bringing dates to the formal dances held every few months. She then closed her remarks by commenting that bringing a date to those events was a *required part of being social*.

We raced again at the Naval Academy to start off the spring season, but sadly didn't win, or even place very well. Back in New London, our practices on the river were spread further apart, as we were already busy with preparatory work for ocean racing. A new boat had materialized from some generous donor, a Farr 40 named *Gem*. We were beginning to dig into *Rampage* and realized with each passing day that she was less and less seaworthy. Holes that we had patched were leaking, fittings were not quite watertight, and much of the deck hardware was worn out. We'd picked up little tricks about how to keep ball bearings intact well past their service life, but the corners we'd cut in the name of previous seasons were all catching up with us.

On the other hand, rumors were swirling about *Gem*. Many of the higher-end boats we'd raced against over the past few seasons were at least partly comprised of carbon-fiber masts, booms, spinnaker poles, etc. We knew little about carbon fiber, other than it was light, which even our thick heads understood to translate to more speed. Most of us barely 21, our minds were consumed with the thought of going faster and beating other boats, leaving little room to worry about the increased risks that might present for an amateurish collegiate sailing team. What we lacked in experience, we more than made up for with bravado. In short, we just wanted a boat that looked cool.

As April rolled around, I wrote to my parents about what I felt to be the increasing frequency of absurd extracurricular lectures we were forced to attend on weeknights. Most of my peers felt the same and our disdain for them was beginning to bother the higher ups. Prior to the evening of April 4th, the corps was instructed as follows:

For the information of the corps,

Uniform for this evening's ops spotlight is SDB Bravos.

The corps is to be seated by 1855 in lower Leamy, as cadets are not allowed in upper Leamy.

The corps is asked to be more professional during the gift giving.

Cheering when the gift is opened is fine, but yelling out, "Give 'em the bird," and other such comments while the speaker is opening the gift is not appropriate.

At each of these events, the guest speaker was often given a miniature model of the *Eagle,* and the corps would cheer wildly when the gift was opened. This had, over time, devolved into cheers of *'Dirty Bird,' 'Love Boat,'* and the aforementioned, *'Give 'em the bird.'* There was also a new thing where the 1,000 or so cadets would stomp their feet wildly on the floor of Leamy

Auditorium in anticipation of the gift giving. To an outside observer, the entire event would seem strange, but it was a small
outlet for most of us, worth a few laughs before we marched
back up to Chase Hall. That some poor mid-grade officer had
to craft the above email speaks volumes about the effort with
which they tried, often in vain, to increase our aggregate level
of maturity even slightly.

In late April, we traveled down to the Naval Academy for the
Kennedy Cup, where once again we raced against the top colleges who fielded offshore teams. We ended up third, which
was not a bad showing, but I wrote to my parents, telling them
rather plainly,

> We screwed up yesterday and didn't sail well. Some
> people lost their edge too and checked out pretty early.
> So from the start we were doomed... we could have won,
> but for some reason, our crew dynamics suck. Our suck-
> iness has more to do with people problems than boat
> speed. It is weird. And it sucks.

With ocean racing in the very near future and the fall season
following shortly after that, I was keenly aware of the demoral-
izing effect of a crew that didn't work well together. I was already formulating a plan to not let personalities get in the way
of winning races. My understanding of sailing was different
than most of the others on the team. In a lot of ways, the sport
was still new to me. Where others had been instructed from an
early age and carried with them some strong opinions of the
right and wrong way to do things, I had somehow survived a
crash course (at times literally) in sailing over just three years.
Perhaps this gave me a different perspective, but I found myself

hesitant to put forth any plan without first calling on the strengths of others around me. I had confidence in myself as I likely had the most experience with *Rampage* of anyone on the team, but I also had a healthy dose of humility to help keep myself in check. I tried to position myself somewhere in between Stan and Brett, whose combined efforts at demonstrating their own styles of leadership trumped most if not everything I'd been told up the hill.

As April came to an end, we got word that *Rampage* would not be put back in the water right away. When she did go back in the water, another crew would sail her. In her stead, we'd be crewing *Gem*, the FARR 40, a boat that no one, including the coaches, had any experience with. She was waiting for us in Newport, Rhode Island and we would be making the drive over to pick her up one weekend and sail her back to New London. In seafaring circles, this is called a shakedown cruise, where the overall goal is to get back to the dock without sinking the boat or anyone getting killed. It's the first opportunity to feel a boat out, identify problems, and begin the process of familiarizing oneself with the sails, rigging, engine, and hull.

Ocean racing was right around the corner, but before that, I had to pass Physics or risk being relegated to *pumping gas at some po-dunk gas station in Hatteras for the rest of my life*, as I so eloquently wrote to my parents. I had managed to pass most of the exams, but not all. I was also magically doing well in the Physics lab portion of the class, but even with that, it was constantly on my mind.

What I hadn't put together at the time was that the physics professor was also one of the safety officers on the sailing team. I don't recall doing exceptionally well, even by my standards, on the final exam, but I ended up with a C- and thus survived another year at the academy. Later that summer, I asked him how it was possible that I had managed a C- instead of the D I was expecting. From the rail on *Gem* where he was seated, he replied with a slight smile that if I had gotten a D, it would have required him to do more paperwork than the C-. I didn't ask any more questions.

With classes over, we had a few weeks before graduation, when I'd officially become a first-class cadet. There would be hurdles during the academic year, of that I was certain, but the bulk of my classes would be within the government major, and I was more confident that I would keep my head above water. From that point on, and for the rest of the ocean racing season, most days started for me around 5:30 a.m., when I'd get up to go for a run, shower, grab some breakfast, and be down at *Gem* a little before 8:00 a.m. We'd work until noon, eat, then work for the rest of the afternoon until 5:00 or 6:00 p.m. Compared to the academic year, I loved almost every minute of it.

Gem was different than *Rampage*. For starters, there wasn't much that needed to be fixed. Rather, we needed to figure out how things worked. Sailboats had come a long way in the 20 years between when *Rampage* was built and *Gem* rolled off the production line. The winches were new, the cleats were all made of lightweight materials, and the rigging was different in that most things were built from carbon fiber. We looked over each fitting and every line on the boat, figuring things out at first in the relative safety of the river. A week or so later, we ventured further south into the Sound to feel her out. On one of our first runs with about 30 knots of wind, we had *Gem* up to 15 knots through the water. I wrote to my parents that I'd only ever seen 14 knots on *Rampage* and that was in the tail end of a tropical storm, the same one where we'd blown out the spinnaker and ended up on our side in the Chesapeake Bay. *Gem* was a fast boat with a lot of potential. Her class was often crewed by semiprofessional sailors, not college-aged cadets. With the speed and agility she offered, there was also an increased risk of things going wrong. Whereas the Luders yawls had been slow, heavy, and stable, a FARR 40 was fast, light, and prone to fits of insta-bility if mishandled. Early in that season, we were all too excited to think about those sorts of things, but the lessons she would teach would come with time.

Chapter 13

This is the press release from the Storm Trysail club themselves. Names such as Gary Jobson should ring a bell with most of you and show that Blue Yankee *was probably the most experienced boat in the whole race and had a crew that was unmatched. Furthermore, the man overboard was nowhere near being inexperienced himself.*

They ran into problems doing a spinny change. We were less than a mile from their location when they lost him, in the same weather system, and had done two spinny changes ourselves as you may remember. Good job on getting them done and keeping relatively cool heads. Our ability on Gem is far beyond what would normally be expected from a rookie crew and will still continue to improve. Don't fail to realize that we routinely sail right at the threshold, just as Blue Yankee was doing when things went wrong. Take some knowledge away from it all and watch out for one another. See you guys in the morning. Be ready to load the boat by 1100. Bring gear to sail, I would anticipate us getting out into the sound and doing a few hours of practice before returning.

AND GET SOME SLEEP!!!!

So went an email to my crew on the evening of 25 May, 2002, after we'd returned to New London without completing the Block Island Race. It had started uneventfully on the afternoon of the 24th, with a modest breeze out of the northwest. With a spinnaker up, we were making great time and settling in as the sun started to fade behind us. I had completed the race twice already during previous summers and knew that we would take the better part of 24 hours to finish. If we could minimize the sail changes early and settle in for a few hours of straightforward sailing, the crew would be fed and better rested in anticipation of whatever challenges we'd face later.

Looking behind us, and with half an hour of twilight remaining, a large dark wall of clouds was beginning to move in our direction. We discussed the best plan for a few minutes, watching as the approaching front gained momentum, and then decided that reducing our sail area before it hit us was the right call. Matt moved quickly. He was short and strong with superhuman balance that enabled him to cover the length of *Gem's* deck in mere seconds. Once on the bow, he led the effort to pull a smaller spinnaker from below, douse the bigger one that was currently full of the now increasing wind, and then clean things up before returning aft.

As he worked, I was back aft, trimming the main and looking over my shoulder at the telltale signs of a cold stiff wind barreling across the length of the Sound. The water behind us was growing dark as the gusts moved closer to *Gem* and drove smaller, tightly spaced whitecaps across what had moments before been a calm Long Island Sound. Even with Matt's hustle, the wind was going to get us before he had a chance to get the sails swapped. *Gem* was fast enough that we were in the front of the fleet, which consisted of 86 boats spread out across the Sound. One by one, we watched the 25-knot gusts hit the boats behind us and knew it was a matter of minutes, if not seconds, before we'd be in the squall.

When it finally hit, *Gem* nearly broached, the bow swinging violently around and into the wind. Matt held on with one hand and continued to work at dousing the spinnaker that was now straining wildly and holding us over on our side. I had dumped the main and was now laser-focused on the crew up forward, watching and waiting, planning our next moves if things continued to deteriorate. Now in total darkness with the sun below the western horizon, our focus was getting sails down and riding out the storm. It was only minutes before we had things back under control. No one was hurt, and the boat was undamaged. We were all certain we'd be back to racing within a few minutes.

It was at that point that we heard the 'man overboard' call on our radio. *Blue Yankee*, which we knew to be one of the top contenders in the New England circuit, had lost someone in the chaos of the storm front. Less than a mile ahead of us, they'd been caught in the same squall minutes after us. Their spinnaker pole had snapped, sending one of their very experienced sailors into the churning water, where he landed face down as *Blue Yankee* sped past in the dark. A second crew member dove in after him.

It took several minutes for *Blue Yankee* to get her sails down and turn around. By the time they doubled back, the second crewmember had lost his grip on the first sailor, Jamie Boeckel, and lost sight of him. After recovering the second crewmember, *Blue Yankee* tried but couldn't find Jamie. We fired up the engine on *Gem* and tried to assist with the search. In time, the squall line passed entirely, and the seas returned to their previous calm. As we searched, someone located the life ring with a light on it that *Blue Yankee* had thrown towards Jamie, and we were certain that he'd be clinging to it, smiling, shivering, and waving as we closed in to pick him up.

The sight of that empty life ring floating in the darkness of Long Island Sound is still burned into my memory. That relatively small body of water now felt immense as I looked back up and around, realizing that my absolute certainty of his recovery was wrong. In minutes, the sea had claimed a life less than a mile from us. It had happened to one of the most experienced boats in the race. Moreover, we'd run the same sail changes that *Blue Yankee* had. I had watched Matt cling to the bow pulpit as *Gem* swung uncontrollably in the gusts. The only difference between our two boats had been *Blue Yankee's* spinnaker pole snapping in two, sending Jamie overboard.

Somewhere in the Atlantic, 2001. Cold, wet, and no doubt tired.

We searched for some time before being called off late in the evening. At that point, we withdrew from the regatta and motored our way east towards New London. When details began to emerge and word spread that one of their crewmembers dove overboard to save his friend, our safety officer told us that none of us would have been permitted to do the same thing. Matt and I looked at each other immediately, both of us trying to hide a smirk. We said nothing, but each of us knew, without any doubt, we too would be in the water if the other went in. It was my first personal experience with death, and I took most of the following day to wrap my head around the previous 24 hours. I sent that email to my crew a little after ten in the evening on the 25th. As I would find out in the coming years, sometimes the best thing—or the only thing—you can do after a tragic accident is get back up in the saddle. The mind won't forget, but in the familiarity of our regular routines, whether it be sailing or flying, you can find comfort and clarity by getting your head back to a new normal.

As we prepped for the next regatta that would be held in Newport, Rhode Island, things continued to deteriorate with Leslie. We talked almost every night, but the conversations were increasingly strained. She finally broke up with me one evening as I was pacing the sidewalks outside of Chase Hall, looking down at the Thames River below. It was hard for her and perhaps even more difficult for me in that I'd been determined not to lose her despite my growing concern that things were going to end. Had I been equipped with more life experience, I may have been able to see the writing on the wall and minimize the pain that we were both going through. But I wasn't at all ready for it and hearing the words come from her mouth cut deep. As bad as I felt for myself, I would most certainly not be the first sailor in recorded history to leave port with a busted heart.

She seemed unsure of what breaking up with me would achieve, but her words were deliberate and there was little doubt in my mind that I could convince her otherwise.

"I don't think we should be together anymore." I could hear in her voice how much she was hurting.

I asked, "Why?" but knew in my head exactly why. I had known for quite some time it was coming, and yet I tried to find ways to undo what had already been said.

"I'm sorry" came next, and the emotion in her voice told me only that she felt as bad, or perhaps worse, than I did.

Next up was the classic, "I don't understand," which was a subconscious lie on my part. I completely understood. I may not have known the exact reasons why she chose that moment, but I knew what she was doing and understood the bigger picture. She and I were two pieces of a puzzle that didn't quite fit then. We'd tried to make our relationship work, and at times they did. When things were good, the feeling was unlike anything I'd ever experienced. But when the cracks began to show, there wasn't much we could do to patch them up.

I wasn't upset with her on the phone; those feelings would come up later. At the end of the call, we said our goodbyes and I wandered back into Chase Hall, my mind running the full spectrum of confused emotions. For a brief second, I felt as if we'd patch things up in a day or two. A breath or two later and I resigned myself to the reality that we were done.

A day or two later, we were back underway and racing in Newport. Thankfully sailing gave me something to focus on other than Leslie. I wrote to my parents that we raced an 18-mile course around Jamestown in winds gusting close to 40 knots. *Gem* was a fast boat, with a small operating envelope and even smaller margins for mistakes. We managed to beat several boats and watched a few broach around us in six-foot seas. With the swells rolling in from the Atlantic, the bow would occasionally dig in hard to an approaching swell and several boats around us lost control, at least one losing its spinnaker pole and spinnaker in the ensuing mess. It was daylight, but not all that different from the conditions that we'd encountered during the Block Island Race the week prior.

Matt and I were both 21, and nights in Newport presented plenty of opportunities for two young cadets to have a good time. We had a regular nightly routine that often resulted in us not finding our way back to *Gem* until the early hours of the morning. On one of those nights, in between establishments, we decided that it was critical to remove the radar reflector from *Gem's* backstay. The small cylindrical tube with reflective material in it was slowing us down, we thought, and if we could shimmy up the backstay and cut it off, we would magically gain some undetermined amount of speed in the next day's races. There was zero doubt in our clouded minds that the coaches would applaud our effort.

We wandered back to *Gem* a bit before midnight and Matt rigged himself up in a crude climbing harness and hooked into the main halyard. Wrapping his arms and legs around the backstay, I proceeded to haul him up the 15 or 20 feet by the halyard. It was a great plan, right up until the halyard snapped under Matt's weight and went shooting straight up to the top of the mast. Matt was able to hold on enough to break his fall on the deck below with only minor injuries. Sans a main halyard, there was no way for us to raise the mainsail for the next day's race.

We made the quick and entirely logical decision to run away. We grabbed a six pack of beer and sat behind a bush in a park on the waterfront, discussing hypothetical scenarios that would play out the following morning. On one end of the spectrum was an understanding safety officer who appreciated our initiative and understood our unfortunate circumstances. On the other end, we took long sips of our beer and speculated that we'd be expelled from the academy and spend the rest of our time on earth trying to piece together our broken lives. The truth, as it almost is, was somewhere in the middle. We slept for a few hours that night and woke early, ensuring our stories matched—we were smart enough to minimize the role played by alcohol and maximize the earnest initiative with which we'd snapped the halyard—and were relieved to find ourselves relatively unscathed. One of the coaches had a friend at a marina just north of Newport who was able to run a new main halyard, and we made the next day's race on time.

Some wise person in a position of leadership had made the smart decision to keep *Gem* close to shore for our first season with her. But another crew had ended up racing *Rampage* to Bermuda and back. I wrote to my parents in late June that *Rampage* had completed the race but returned to New London with the top third of her mast missing after a storm just south of Montauk, New York. Making matters worse, the engine had crapped out once again, requiring a tow from a Coast Guard Cutter. This was likely the *coup de grâce* for the old boat, and I don't believe she sailed competitively again after that.

Matt and I race during my second-class summer, somewhere in the Atlantic.

At the time, I was somewhat thankful to not be stuck with a leaking boat, yet I held nothing but fond memories of my time aboard *Rampage*. It was where I had learned the ropes. She had given me my first experience on open water—on her deck I'd felt the waves and warm water of the Gulf Stream, navigated fog banks in the pitch black, avoided commercial tankers at the last minute, and found that I possessed the strength to power through endless wet and cold nights at sea. We'd battled storms, and I'd hid whatever fear I felt for no other reason than to not let Stan down. I'd seen his unparalleled leadership shine and watched others fail. *Rampage* was perhaps the most perfect leadership laboratory that the Coast Guard Academy could have assembled. Yet those lessons were entirely unscripted. There was no textbook or series of lectures; our instructions were little more than to sail or race from one port to another. The ocean was our classroom. And I'd been there to see it all.

Our summer season on *Gem* ended with the Block Island Race Week, which was a familiar event that I'd sailed in for the past two summers. Day races were usually completed an hour or two before sunset, the small and exclusive town then coming alive to provide a festive and rowdy place where I could cap off three seasons of ocean racing. I also convinced our safety officer that during a few days with no scheduled sailing, it would be in our best interest to make a run over to Martha's Vineyard in the name of proficiency.

It was one of a handful of times where we were not under the pressure of a race, or the time crunch of prepping for a race. Matt and I harnessed as much as we could remember from Nautical Science and planned a route out the mouth of the Great Salt Pond, then around the island before running east across Rhode Island Sound and north of Martha's Vineyard. From there, we initially planned to overnight in the harbor of

Vineyard Haven. The morning of our departure was perfect summer weather with a healthy breeze from the southwest. Rounding Block Island, we threw up a spinnaker and met rolling swells coming in from the Atlantic as we ran downwind towards Martha's Vineyard. I took a turn at the helm and spent much of the run across the Sound bringing *Gem* up to a beam reach, taking her hull up to the crest of a swell, then down to a broad reach and riding the wave while *Gem* burned off the energy she'd built. Our safety officer wasn't thrilled with my antics, but we reached speeds of 17 knots, and my face hurt from the big smile and pure bliss of surfing with a nearly 11,000-pound sailboat.

Once in the lee of Martha's Vineyard, we doused the sails and motored towards the entrance of Vineyard Haven. I was scouring the sailing guide as there were potential shoals near the entrance. One single line caught my attention when I read that Vineyard Haven was a dry town. I quickly summoned Matt to look and make sure I hadn't misread it. He confirmed my fears, and we quickly concocted a story that would prevent our continued passage into the harbor. I had the helmsman turn *Gem* around and directed us towards Edgartown, just a few miles further east. Our safety officer was quickly up on deck, wondering what all the commotion was. I blurted something out about dangerous shoals and our draft being too deep, which Matt then backed up to solidify our story.

Thankfully, our officer didn't ask too many questions, and we made an uneventful course correction to our original plan. Once anchored in Edgartown, and with *Gem* put to bed, we made our way into town. I had made some kind of arrangement with a friend of a friend who lived on the island, and Matt and I soon were tooling around Martha's Vineyard with a rather attractive girl home from college for the summer. We spent the rest of the day at the beach, then at some other equally attractive girl's pool, and then finally back to the boat that night to get some sleep. It was far from the wildest night of my ocean racing experiences, but still a fitting end to the season.

Back at the academy, I had orders to Portland, Maine, where I was to spend a month on a 110-foot patrol boat. I had managed to miss the otherwise obligatory ten weeks at sea that most of my classmates endured that summer. I had also skipped the five weeks afloat during my third-class summer, all on account of ocean racing. I wasn't entirely sure of what to expect when I arrived. Loading up my new Toyota Tacoma, I drove north and found my way to Portland, where I reported aboard the Coast Guard Cutter *Jefferson Island*.

The only officers were the commanding officer and executive officer, and the remaining crew of 16 were enlisted guys. Both the CO and XO were married and had little interest in showing me around or being social after work hours. I was aware that the enlisted men didn't trust cadets and was relieved when, on my first night, the crew invited me out to celebrate someone's going away. Portland had lots of bars, many of which this crew knew well. I immediately felt at ease as we rambled around downtown from bar to bar. Towards the end of the evening, I had attracted the attention of a lovely young woman with long blond hair and a pierced nose. She and I left the last bar together and this earned me immense street credibility with the crew, and for the next four weeks, they considered me one of their own.

The CO and XO were good enough to let me feel things out. Normally during a third-class summer, the intent is for a cadet to experience the day-to-day work of the enlisted crew. Then, during one's first-class summer, that same cadet is expected to shadow the officers and learn the particulars of their respective jobs more closely. In my case, this was my first time on a cutter, and, just as I'd experienced at the waterfront, I was most comfortable working side by side on the deck, rather than from the confines of my stateroom.

A patrol boat was also an interesting case study in the art of dealing with people. The crew wasn't much larger than what I'd seen on *Rampage* and *Gem*. Each member had their own little quirks, their own strengths, and their weaknesses. When totaled up, they could take that cutter anywhere in the western hemisphere. I watched them work together to get the boat underway, navigate the congested harbor, and let the engines loose once we hit the open water. Whatever differences they'd had ashore—and there were plenty—were cast aside when they got down to business. The CO and XO were at ease as well, keeping some distance from the particulars of how any evolution was accomplished. If anything, they offered subtle coaching, little more than gentle nudges to keep the ship and crew headed in the right direction. To me, it seemed no different than racing sailboats, and I enjoyed my time with them immensely.

As I would find out the summer after graduation, not all cutters are cut from the same cloth, and my intention to spend my career at sea was short-lived. But I left Portland quite happy with the experience. The ease with which the CO and XO handled the *Jefferson Island* was somewhat of a façade. There were personnel issues that had to be handled, and discipline likely enforced out of sight. As I'd discovered with sailing, one bad personality can ruin the cohesiveness of a small crew. I was fortunate to see and experience some good leadership from those two on *Jefferson Island*. I can only hope that they got a bit of a chuckle from watching me slowly figure things out on my own.

And perhaps they appreciated the two cases of beer I left dockside for them on my departure.

Chapter 14

I had a few weeks off in August before the beginning of the fall semester. Home in Virginia Beach, and now without a girlfriend, I took the money I'd been saving in the ever-increasingly unlikely event that I might one day buy an engagement ring and blew it all one morning on a pretty awesome mountain bike. In between that and surfing, I had a decent few weeks before making the drive back to New London in my still new-smelling truck. There was little room for pity, but I was still reeling a bit from the breakup with Leslie.

In late August, I once again walked into Chase Hall and got on with life. Within a week, the fall sailing season was in full swing with a late August weekend spent in Newport at the East Coast Championships, where we'd be up against other Farr 40s crewed by career sailors. The summer had given us some good experience with *Gem*, but any weaknesses that we had not yet ironed out would be on full display against of fleet of similar boats with professional crews. The season would run through late October, with the McMillan Cup down in Annapolis as a capstone event. My crew was not the same that I'd sailed with during the summer, most notably having lost Matt back to the dinghy team. But I knew the boat well enough to try my hand at teaching the new faces. Perhaps more challenging would be the task of learning individual strengths and weaknesses. I saw my job as team captain as one where I'd be ultimately judged by how well I could get them all to work together. We could get the boat around a racecourse, but if we could maintain some cohesion and build a good foundation, the boat would sail faster, our evolutions would run quicker, and perhaps we could hold our own against the more experienced boats.

Walking the halls during that first week, it was apparent that a fair number of my classmates had let their summer cruises—and the many port calls—get the best of them. I wrote to my parents speculating that many were likely going to struggle on the fitness test. I wasn't too worried, as I'd kept up a good routine of running and working out all summer. Grades were always on my mind, and I would be damned if insufficient pushups were the reason I found myself escorted out through the front gate. Some, however, had let the impending fitness test slip their minds entirely. I ended up with a score of 408 out of 500, which was 42 points short of the blue star, and I had resigned myself to never making the "Athletic Director's List." I could run the mile and a half—in 8:52 as a matter of fact—max out push-ups, pull-ups, and sit-ups, but my genetic makeup forever precluded any kind of respectable performance on the shuttle run and long jump. Years later, I read more about those fast-twitch and slow-twitch muscles, and how the human body is hard-wired in a certain way at birth. I took this a step further and reckoned that my brain was similar in that I never seemed to pick things up at the first go, but I also wouldn't quit if it really mattered. *Slow and steady wins the race,* I assured myself time and again. I was the proverbial ornery tortoise in a field of energetic rabbits. This mantra would hold true throughout my career.

Throughout my flying career, I felt a little slow out of the gate when learning a new plane, but I never entirely put down the books and rested on my laurels. I watched a fair number of pilots struggle both in flight school and at operational units who did just that.

Academically, I was feeling good as most of my classes were within my major. I was also taking Spanish I, as two semesters of a language were required. Towards the end of August, the teacher was trying to push me from Spanish I into Spanish II. I had retained enough from my Spanish classes in high school, and he felt that Spanish I wasn't pushing me enough. I fought back hard on that one, realizing that a relatively easy language course would free up enough time to focus on Intro to Electrical Engineering, which even though it was 'introductory,' still scared the living hell out of me. I asked my folks if they could make any sense of expressing the *following time domain function in its frequency domain (phasor) form: v(t) = 120 sin(4t+pi/2) > V(jw) =.* They could not, and neither could I. Within a week of beginning the class, I scratched 'electrical engineer' off my list of potential things I wanted to be when I grew up. Luckily, my Spanish professor relented and let me stay, giving me a chance of offsetting whatever damage was in store from what we all affectionately referred to as *IEE.*

Our sailing season was more intense than previous years. We were travelling a lot more with *Gem,* and the program was putting us up against professional crews in regattas that were forcing us to fight well above our weight class. We finished last at the East Coast Championships. By September, I was frustrated by the schedule and by the day-to-day business of being the team captain. The title of 'Team Captain' had seemed much cooler the previous summer when I didn't actually have to do anything to earn it. Often, we would transit a day early, almost always leaving in the afternoon and arriving at whatever yacht club late at night, only to be driven back to the academy before leaving early the following day. The fun that we'd had during ocean racing was non-existent as the coaches were trying to squeeze as much from us as they could.

What I failed to realize at the time was that, with each regatta, even if we ended up low in the rankings, we were, as a crew, pushing ourselves harder than we ever had during past seasons with *Rampage* or *Steamboat*. A highlight was thumbing through the monthly sailing magazines and seeing pictures of our regattas, at times even spotting *Gem* tucked in somewhere among the fleet of FARR 40s. The pressure was exponentially higher on *Gem*, as any glitches in the sail changes or evolutions that we ran during a race were going to put us into an embarrassing tail position. We were not winning races, but we were consistently somewhere in the pack, a pack of professional sailors who'd been doing this far longer than any of us had.

My biggest challenge was not teaching anyone how to hoist a spinnaker or trim a jib. It was finding a way to keep my crew motivated enough to give it their all for each subsequent race when we all knew it was unlikely that we'd ever cross the finish line first. While not a deliberate act on the part of the coaches, they were giving me the opportunity to put what I'd learned during previous seasons to practical use. I channeled my inner Stan and then harkened back to the humility I'd seen from Brett, trying my best to take it all in stride. The tone of my emails to my parents was generally *I'm so tired of this crap*. My dad sent me one of his many sage replies, in which he told me:

> "One thing USCGA has already enabled you to earn is the precious experience that lets you know you can do difficult things under difficult circumstances. When I reported aboard Gitmo (as the commanding officer of the Base) I thought to myself 'they've got to be kidding.' But I also was able to truthfully say to myself "I've seen & handled worse stuff than this." You're the same. So press on my son. There are very good things close at hand for you that you have earned."

This was, of course, the kind of timeless advice that would benefit anyone in a position such as mine. However, given my age and frustration, it calmed me down for the better part of half an hour before I let the same old frustration boil up inside me again. The end was in sight, there was no denying it, but with the fall semester not yet halfway over, the light at the end of the tunnel seemed to flicker and fade almost every day. Winning races would have been fun, but that was not in the cards either, at least not while we were on *Gem*. We were destined to run the remainder of the season without ever claiming a race for our own. Whatever progress we made paled in comparison to the more experienced crews, and it was increasingly disheartening.

In mid-September, I revealed a devious plan to my parents that I'd been concocting for some time. I reckoned that with my experience, I could make a run for France in a J/22 and be at least two days out to sea before anyone at the academy realized a boat was missing. My plan included water, granola bars, a compass (of course), and charts. While my provisioning was probably lacking, I knew the routine at the waterfront enough to be certain that it would take at least a day for anyone to realize a boat was not in its slip. After that, confusion would give me another day or two before anyone realized they were short one hull. After that, the search would primarily focus on yacht clubs and marinas along the eastern seaboard. No one, I figured, would even think that I'd made a run due east for foreign shores and pretty girls with curly hair and French accents.

I closed the email outlining this plan by assuring my parents that with graduation looming and my truck in the cadet parking lot, I would defer going AWOL pending further developments in my academic status.

Introduction to Electrical Engineering continued to give me trouble. At some point, the professor remarked that the first exam had been too easy and that there would be no curve in the class for final grades. This was not good news for me as I'd only managed a 58 on the test. I had a feeling that the odds were in my favor. I was a first-class cadet, and by that point, Uncle Sam had invested a fair bit of money into me. On top of that, I was a government major forced into a class I didn't want to take. Surely there would be some mercy bestowed upon me at the end of the semester if I pledged to never mess with electrons again in my life.

Maritime History, Maritime Law Enforcement, and Spanish were easy enough for me to hold a decent grade and only worrying about one class, electrical engineering, made life more bearable. Oceanography sounded cool, but as I found out it had more to do with science and math and less to do with surfing and sailing. Still, I was holding my own in that class as well. On top of that, I was taking a scuba elective and halfway through the semester, we were cut loose at a local beach to dive in pairs. The instructor was a surfer and he and I got along well. He would often let me do my own thing in the water. The intent of most of the class was to familiarize students with basic swimming techniques, how the equipment worked, and being comfortable in the water. A few of us had taken to wrestling at the bottom of the deep end, and we had a blast trying to rip goggles and masks from each other's faces.

For our open-water dives, I would routinely bail from the group and make my way out a bit farther and into deeper water, usually managing to find my way back before the herd resurfaced. Being immersed in the salt water of Long Island Sound was like jumping in the Thames during my Swab Summer. It was cathartic to dip under the water and swim my way out to explore, even for a short time. Once, I was late, and when I surfaced several minutes after the rest of my class, the instructor was laughing and shaking his head. He got innumerable cool points in my book for not reprimanding me.

At the end of September, we raced in the Shields Cup at the Naval Academy and placed third. It wasn't a bad showing, but I replayed the events in my head, trying to think of ways for us to do better at the upcoming MacMillan Cup. As I saw it, there were three elements that needed to come together to win a big regatta. The first was tactics. Someone on the crew had to understand not only the rules of racing, but also have a keen eye to understand the ever-changing dynamics of a racecourse. Dinghy sailors, in my opinion, were the greatest at tactics, within the limits of their smaller boats. The second piece of the puzzle was a crew that could tack and jibe the boat, change sails, and constructively react when things were on the verge of falling apart. The last part was the glue that would hold those two pieces together. This was where I could thrive. If I could put a team together for the MacMillan Cup that pulled from both the offshore team and the dinghy team, we'd have the best of both worlds. The challenge would be getting everyone on the same page over the span of a few days of practice on the river. I had the third-place trophy from the Shields Cup in my room and stared at it, sometimes for quite a while, and spent what little free time I had putting the finishing touches on my plan to win the MacMillan Cup.

In mid-October, I wrote:

Some guys are threatening to quit and others are throwing tantrums about stupid stuff. This last regatta at navy is looking like crap right now. One guy wants to quit, one guy thinks that guy sucks and should be kicked off. Then another thinks the guy that thinks someone should be kicked off has too much of an ego. Blah Blah Blah. And the worst part is they all come to my room explaining why they want to quit or why someone should be kicked off the team or who has an ego and who doesn't.

Most of the ones I had problems with were my classmates. At this point, we were all firsties and wrongly thought that we were somehow important or a big deal anywhere outside of Chase Hall. The bickering that I encountered leading up to the MacMillan Cup was a phenomenon that I would encounter again, and frequently throughout my career in the Coast Guard. Put a group of people together, most if not all of them higher-than-average performers, and in the absence of a clear rank structure, infighting would ensue until a pecking order was established. I was the team captain, and each of them tried to convince me that their way of doing things was the right way, hoping to establish themselves as superior to the rest of the team. On paper, I was in charge, but in reality I had a fairly narrow swim lane to stay in.

My first push with the coaches was to bring in two sailors from the dinghy team. Matt and Chris were the names I specifically requested. I had a great relationship with Matt and knew that Chris, in addition to being an accomplished dinghy sailor, kept a pretty level head in all our previous encounters. I needed one of them to drive and the other to call tactics. I was confident that Chris and Matt would work together seamlessly. There was some hesitation to pull them from the busy dinghy schedule, but eventually the coaches relented, and I got my way.

The rest of the crew would come from the offshore team. I had sailed on several boats with all of them and had a good idea of what they individually brought to the table. The crew was top-heavy, with the majority being classmates of mine. This presented the biggest hurdle, as some of them no doubt thought they had better ideas than me. We practiced a few times out on the river, pushing right up to the brink of a meltdown among the team. We could get a boat around the buoys, but if the crew wasn't going to mesh well, things were destined to fall apart. In the final week or two before leaving for Annapolis, I wasn't entirely certain we'd make it through the whole regatta without some significantly bruised egos—or eyes.

Driving down to Annapolis the night prior, after settling into a hotel, we all got dinner as I anxiously counted down the hours. Down at Santee Basin the following morning, we prepped the boat assigned to us, double- and triple-checked everything, then motored out for the start. Most of us had some experience with the Navy 44s and were familiar with the rigging. Chris and Matt had to quickly get acquainted with a very heavy hull. The snap tacks and jibes would be a lot slower than they were used to. Each evolution posed the risk of a line run incorrectly or a sail not trimmed in time that would push us back a few spots in the pack. Recovering from those errors would take precious time.

I trimmed the mainsail for the regatta but knew that my focus would be on running interference between both ends of the boat. Matt and Chris could read the racecourse and knew what they wanted the boat to do. My job was to translate that into something the rest of the crew could understand while also tempering any unrealistic expectations from either end. Halfway through the first day, we were holding our own. While we didn't know exactly where we were in the pack, we all sensed that we were near the top. In between one of the races, things began to break down. Someone was apparently at their breaking point. Deep down, I suspected that he was upset at having not been given the opportunity to call the shots. I was certain that he was not the best guy for the job, but he was certain that he was. The Naval Academy provided an observer for each boat who also acted as a safety officer but offered nothing in terms of coaching or advice. I was on my own. With perhaps five or ten minutes before the start of the next race, all eyes were on me.

I may have done a poor job of dissuading the offended crewmember from escalating his temper tantrum as I knew that he was digging himself deeper with each insult that came out of his mouth. But as the minutes ticked down, I realized that I needed to do something. At about that time, he threatened to quit and the whole boat went silent. I looked at the safety officer who was looking directly back at me with a blank look on his face. In two or three minutes we needed to be racing again. I told him if he wanted to quit, he was welcomed to, but would have to swim back to shore. I added some emphasis by pointing back towards the Naval Academy, perhaps a half-mile or so to the west. This got some laughs from the crew and our rogue sailor, now defeated, quieted down. We regained our composure and made the next start on time.

It may not have been the best way to handle things. I doubt public ridicule had been one of the methods taught to us as a means of diffusing tension. In a perfect world, we could have all sat around and talked about feelings while doing trust-falls or something equally moronic. But facing a limited amount of time, I had tried to leverage a little bit of humor to de-escalate a tense and toxic situation. He'd already made a fool out of himself and was not likely going to listen if the two of us engaged in an escalatory shouting match with each other. His anger was directed at Matt and Chris, and I needed to deflect that away from the two of them. Bringing them along was my idea, and I needed to own it. It was clearly working for us, and thankfully most of the crew was onboard with the plan. We finished the Saturday races without any major problems.

Sunday then gave us a good second day of racing. It would be a stretch to say that we'd patched up all the hurt feelings, but we did hang out as a crew. It was equally important not to dwell on the past. Thankfully we finished Saturday near the top of the fleet, and this gave us much-needed motivation going into day two. With some momentum on our side, we continued to win races. The pace with which we'd run *Gem* that fall paid dividends as we dialed in the Navy 44 and fought off the tactics of the other boats. By the final race, we were a well-oiled machine as we crossed the finish line.

Dropping the sails, we passed the committee boat with the coaches onboard. I caught a smile on the face of our coach as he gave us a thumbs up. Looking at his notes, then back at us, his grin kept growing, and I knew that we'd won the Cup. I realized that our success was my biggest accomplishment. I'd walked on to the sailing team knowing little about racing sailboats. I'd survived more than three years of teachers and advisors looking down on me with scorn. I'd felt out of place for as long as I'd been a cadet. Yet somehow that win instilled the much-needed confidence that I'd been so hard pressed to find on my own.

Walking around the sailing center, I held my head a little bit higher. My parents had made the drive up to spend the weekend in Annapolis, so they got to see their son in a good spotlight for once. I'd written a novel's worth of frustrated emails back to them over the past few years, and there can be no doubt they were relieved to see what I'd been able to accomplish. We were presented with the trophy and quickly loaded up into a van for the drive back to New London. I slept, the MacMillan Cup resting by my side as we sped north into the night up I-95.

Chapter 15

As November rolled around, the sailing season was over, and I had more free time. Increasingly worried about Intro to Electrical Engineering, I was disheartened that Maritime Law Enforcement was proving to be more difficult than I had first thought. With IEE, I was hopelessly lost, but I had expected to do better with law enforcement. As a branch of the military with federal law enforcement authority, the Coast Guard was routinely called upon to enforce both domestic and international laws on the water. We were thus expected to be familiar with the complicated framework of both U.S. and international law to pass the class. I was not particularly fond of the methodical research required to figure out the legality of the many high-seas scenarios our professor threw at us. My approach was very much a "shoot first, ask questions later" mindset which was earning me a D as the semester neared its end.

I had hoped that this year would be devoid of the day-to-day academic problems encountered during my first three years, but this was not to be the case. I continued to watch classmates of mine, now less than six months from graduating, being expelled for major infractions that tended to have an alcohol component to them. There were occasionally events at night where the academy would allow us to indulge in some adult beverages. Some were formal, like a night dedicated to lectures about the cutter fleet, where officers from around the fleet tried to extoll the virtues of their various cutters and get us excited about our upcoming first assignment at sea. Others were purely recreational, such as the Monday-night football games broadcast in the officer's club where first-class cadets could watch the game and drink beer. I never attended any of them, as I'd seen enough of my peers in serious trouble for the rest of their time at the academy due to poor judgment at the tail end of an evening. I wasn't perfect—far from it. But I did learn early on the real-world skill of managing risk. Away from the academy, among a small and trusted group, there were times we all let our guard down. But walking back to Chase Hall intoxicated on a weeknight, with officers lurking around each corner and mischief on one's mind, was never a good idea.

I spent most of my scant free time one state over in Rhode Island. Armed now with a truck and a few surfboards, I had explored the coast and knew half-a-dozen good spots to catch any northerly swells that rolled through and a handful of good places to grab a beer and burger on the drive home. I was magically not on any academic probation and free most weekends to roam the coast in search of surf. It was unintentional, but setting out by myself also kept me clear of the kinds of trouble that seemed to follow larger groups of my peers who were now just barely over the legal drinking age.

As December rolled around, my class was down to just over 200 cadets remaining from the 326 that had reported aboard a little over three years ago. We were nearing 40 percent attrition, and more would follow before our scheduled graduation on May 21st, 2003. My only real threat: I needed to pass IEE. A week or so before final exams, I wrote to my parents complaining about having to go into Macallister Hall to deal with the *engineers and their lifeless bodies walking around in circles with calculators and graphing paper and equations and coffee breath.*

IEE being the only threat did not mean that I wasn't busy. Leading up to finals, I described a typical evening routine, where I attended *mandatory alcohol training at 1900 until*

2000. At 2000, I went to an IEE study session for an hour. Then at 2100, I had to get back to my room to write a summary paper for Maritime History. Once I finished that up, I studied for Spanish. And once that was over, I read over the entire scuba manual to make sure I was fresh on all that stuff for the final exam the following morning.

The work paid off, and I managed to close out the semester with passing grades and no real damage to my class standing. I somehow managed to pass IEE with a 61.5% score on the final, earning the coveted D I had been hoping for. I ended up with a 2.33 for the semester and a cumulative GPA of 2.23. I spent a week or so at home for the holidays before flying out to San Diego to stay with a friend and get some surfing in. From there, I packed up again and caught a flight back to Providence for what I hoped would be my final semester. My birthday happened to fall on a holiday weekend shortly after our return to New London, and I took off with some friends to snowboard in Vermont. We ended up at a one-star hotel with green goo dripping out from underneath some cracks in the shower, but it was a fantastic weekend away from New London. Driving back, I got in late and climbed into my rack exhausted and cautiously optimistic about the next few months.

Just as the semester was set to begin, more heads were prepped to roll as another scandalous event boiled to the surface. A classmate of mine who was dating a girl from the class under us—which was completely on the up and up—happened to also be dating several other girls from the class below hers—which was not. Dating anyone more than a year away from your class was expressly forbidden. It might not have made headlines had it not been for his understandably upset first girlfriend, who took it upon herself to email not just my entire class, but also her class, meaning that there was absolutely zero chance of his misdeeds not making their way up the chain. The class of 2003 would soon find itself numbering less than 200.

On the docket for my final semester, I was looking at Spanish II, Contemporary Foreign Policy, Political Development in Sub-Saharan Africa, Democracy in America, Martial Arts, and Nautical Science IV. We also had to take the Deck Watch Officer exam once again, the same one I'd already passed. The academy felt it appropriate to make us take it a second time in case anyone had forgotten the lighting requirements for a barge under tow whose length exceeds 200 meters. (If you're curious, it's three masthead lights in a vertical line, sidelights, a sternlight, and a towing light in a vertical line above the sternlight.) Nautical Science kept me the busiest, but it was not nearly as menacing as the engineering, math, and science classes that had dogged me for three and a half long years.

In between the increased liberty, I continued to run the circuitous routes I'd mapped out over the past few years after classes on most days. The short winter days meant that the sun was down before five in the afternoon, so most days I slipped out right after class and ran in below-freezing temperatures with whatever little bit of fading orange light slipped between those ever-present heavy and low New England clouds. Otherwise, my surroundings were reduced to shades of white and black. Constant snow, hulking grey rocks, and a dark river running south dominated the landscape.

My martial arts class proved to be a surprise hit. I had no prior training, nor did anyone else in my class, but our instructor seemed to take it seriously. We learned various sequences of moves to both attack and defend via traditional Tai Kwan Do or something similar. All in all, it was a low-threat class with a high probability of a decent grade. What really made it special was the sparring towards the end of each class. With gloves and some pads for our protection, we'd square off with a partner, bow, then when the buzzer went off, we'd proceed to try and beat the snot out of each other. I didn't mind getting hit and don't recall taking too many hard blows, but I loved going toe-to-toe with some of my classmates that I was not too fond of. Whatever frustration I'd been harboring for the past few hours quickly found its way out via my fists.

With the buzzer sounding a second time, we'd bow again—me often with a devilish grin on my face—then switch partners and go at it again. At the end, we were all soaked in sweat, snot, and a little blood from our noses or lips. The class was right before lunch, and I was often in a moderate amount of pain walking back up towards Chase Hall, at the same time reeling from the strange catharsis that comes from using one's energy to affect physical pain on someone else.

Spring Break rolled around in early March, and I made my way down to Costa Rica with some classmates to surf. We ended up at a primitive camp on the Pacific coast, and it was a fantastic week of warm water, cold beer, and pretty good waves. I sent my parents some pictures from the surf. I made it back to New London in mid-March to begin the grind through the remaining two months. The spring sailing season was set to begin the following week.

We were back on the J/22s, brushing snow off the deck most afternoons and racing around a small triangular course out on the river. Our numb fingers and toes often signaled the end of each practice and yet another walk back up the hill for a hot shower and some food. It being my final season, I can't say that I minded it all that much. We were prepping for the Shield's Cup at the Naval Academy, which was at that point only a few weeks away.

Mid-terms came and went. I was caught off-guard by the lack of emails and letters in my inbox telling me of my sub-par academic performance. I managed a 2.6 and was holding myself on solid ground as the end was now truly in sight. I'd also gotten orders to a 270' cutter, the *Tahoma*, which was based in New Bedford, Massachusetts. It was slated to change homeports, but at the time I was unsure of her destination. A commander had told me that he was certain it would end up in Newport, Rhode Island, which got me excited. Weeks later, I learned that it was in fact going to move to Portsmouth, New Hampshire. This was not all that bad, but I did learn a valuable lesson never to believe anything a senior officer tells you when he or she insists they are 'absolutely certain' about anything.

We raced for the Shields Trophy in early April and placed somewhere in the middle of the pack. It was match racing, where two boats go head to head trying to both outsail and out-maneuver each other. With only two boats, there is a fair bit more involved in calling tactics as the two teams are dueling against each other. While I could tack and jibe and change sails, I still wasn't—nor would I ever be—an expert at tactics. We had two teams competing, and neither of us were able to get ahead at the regatta. Still, it was an enjoyable experience to be racing one last time at the Naval Academy. With graduation around the corner, I found myself introspective about the whole cadet experience that was soon ending.

While I hoped to continue sailing in the future, I preferred that it be aboard boats that permitted a few cold beers on the transits and after races. When aboard academy boats, it had been an entirely dry affair, but I'd seen that it was quite possible to sail at an easy seven or eight knots with a beer in hand and avoid catastrophe.

On Wednesday, May 7th, I took my final exam in college. It was Nautical Science, I ended up with a C+ for the class, and I felt that summed up my experience at the Coast Guard Academy. I had thought for some time about what I would do after my last class ended. Often, I imagined myself running outside and throwing my hands over my head, screaming at the top of my lungs that I had finally made it. Instead, I took a casual walk back to my room, where I changed into some gym clothes, and went for a run around the campus. After that, I cleaned up, headed for my truck, and took off towards Rhode Island to find some surf for the rest of the afternoon. Anticlimactic as it may have been, it was still an unforgettable feeling.

Between my last exam and graduation week, I had a week of intelligence-officer training to attend in Yorktown, Virginia. It was a very low-threat week where Matt and I shared a spartan room on base with little to do after classes besides grabbing a case of beer and sitting around, ruminating about the past four years. At the end of the week, we were now ever-so-slightly more qualified to assume the collateral duty of 'Command Intelligence Officer' once aboard our cutters. I drove back up to New London with graduation week the only thing between myself and a commission.

There were a handful of events to attend and likely a few more parades for the same parents that had tearfully said goodbye to their kids four years earlier. At an awards event, I was named the cadet that 'excels at theoretical and practical seamanship,' which could only be attributed to the sailing coaches having a deciding vote on that particular award. There was absolutely no chance that my below average grades in Nautical Science had anything to do with it, nor did any of my teachers from the past four years. Nevertheless, I walked up on stage and was given a nice pair of binoculars with my name engraved on the carrying case. It was slightly embarrassing and gave my parents a good laugh.

Graduation itself was an unremarkable event. In true Connecticut fashion, it was a grey and overcast day with a sort of hybrid fog mixed with rain throughout the event. Sitting in my seat, in dress whites, I shivered and wasn't sure if I was cold or if it was nerves getting to me. I walked across the stage and was presented with a degree and a commission, the former being given to me by President George W. Bush, and the latter given to me by my dad, who had donned his own dress whites for the occasion. I then found out that the rolled-up degree and commission were in fact blank, and we had to wait in a line later that afternoon to get our real documents. My degree was held up for a few additional minutes until I settled an outstanding debt with the library for a book they claimed that I never returned. I recall thinking that writing a check for a few dollars to them was a pretty small price to pay for the previous four years.

I had orders to Coast Guard Cutter *Tahoma*, but before reporting aboard I had almost 30 days of leave to spend any way I saw fit. In between leave and my first assignment, I'd negotiated a deal where I would report back to the academy for approximately 60 days to help the waterfront with the newly reported Swabs who would make up the class of 2007. I had very little clue what lay ahead, and while graduating was an accomplishment to be proud of, the day seemed insignificant when I looked back on my time there. I didn't feel prepared. The journey spanned from July 6th, 1999 through May 21st, 2003. I had learned quite a bit, but I struggled to define exactly what those lessons were.

In hindsight, the academy experience was a journey where we had all stumbled on more than one occasion. I was not surprised by the struggles I'd encountered along the way; I knew going into it that the experience would be difficult. But I was not prepared for the frequency of that relentless onslaught. I felt as if I'd survived rather than thrived and wondered if any of my classmates had the same feeling. So many of us put on a brave face when others were watching, but I remembered how I'd felt at my weakest moments. I had vowed not to quit, and I hadn't. At the same time, I hadn't fared nearly as well as others. The ones that had quit or been expelled, nearly half of those who had reported in on that July afternoon in 1999, were nothing more than an afterthought. It was the ones who had graduated that mattered. We were all headed to the fleet, and if academic performance was any indicator of future potential, I still felt a fair bit of insecurity.

What I didn't understand was that those feelings were simply a manifestation of the inherent need to continue to learn and grow. Graduation did not mark one's mastery of leadership or nautical acumen. Rather, we'd been equipped with the bare essentials to begin yet another journey where there would be more hurdles and missteps as we gained experience as officers in the fleet.

The academy had done a remarkable job of breaking down whatever misplaced confidence I'd had prior to reporting aboard. I'd been reduced to a blank slate, which was by design. The chaos of the Swab Summer experience was reinforced with that first academic year as a fourth-class cadet. Teamwork was stressed and often a vital component to our survival. But each of us fought an individual battle to stay focused on the long game. Perhaps for some it came easy, but this was certainly not the case for me.

The process of building us back up into the future officers the academy hoped we'd become was a far less defined path. I would argue that no real curriculum existed for the officers and civilians that put their efforts into doing that. I learned just as much from those I hoped to never emulate as I did from those that I admired. It's fair to say that as a class, we saw both the good and the bad of what the Coast Guard had to offer, which may come across as a harsh criticism, but in fairness even the highest-ranking officers were still learning alongside us. The academy, in that sense, acts as an educational pressure cooker where people's best—and at times worst—will be seen, judged, dissected, graded, and discussed. If a cadet pays any attention while in the thick of it, they will come out on the other end with a profound education born not only from the classroom, but as a sum of the experience for those fortunate few given the opportunity.

Chapter 16

My first assignment was not at all what I expected. Life as a junior officer on a cutter had much more to do with following instructions than it did with seamanship. We were underway often, anywhere from the North Atlantic down to the southern Caribbean Sea. I qualified as a Deck Watch Officer, which gave me the opportunity to conn the ship when we were out at sea, but most of the junior officers, including me, were never given the opportunity to actually conn

Onboard CGC *TAHOMA*, circa 2004

the cutter in confined waters or when mooring up to a pier. Even at sea, any deviations from the very strict guidance given to us by the commanding officer required a call down to the captain's cabin to seek permission to alter the ship's course or speed.

The stakes were obviously much larger, with a crew of around 100 and a ship that weighed several thousand tons, but I became frustrated by the lack of opportunities to demonstrate any of what I'd learned over the previous four years. The cutter community prided itself on ensuring that the new junior officers were made aware of how difficult life at sea could be, or in their view, should be. At times, they sought out opportunities to make things difficult to teach and mentor us impressionable young officers who were so clearly in need of tough love. I had less and less patience for this cutter-concocted crucible that we were expected to endure for the entirety of our first tour. Many of my peers, who in fairness were far more intelligent than me, found ways to cope and endure while staying in the good graces of the command. I, on the other hand, did a horrible job of masking my displeasure, and it was made perfectly clear early on that I was far from the favorite child.

This situation didn't mean that I never found opportunities to apply some of the things I'd learned on the decks of *Rampage* and *Gem*. During the first fall aboard *Tahoma*, we were steaming several hundred miles out in the Gulf of Maine when I spotted a sailboat with its sails luffing in a steady breeze and moderate seas. The sheets for the jib were being blown around, and the boom was swinging from side to side. I discussed it for a few moments with the Officer of the Deck, Mike, who was a good friend of mine. It was the first instance where I unwittingly called on my sailing experience to ascertain that something was amiss. We both agreed that it was unusual to see a boat drifting with no one deck so far out to sea. Mike then called down to the captain to report what we'd seen. Moments later the captain, executive officer, and operations officer were all up the bridge. I tried my best to explain my gut instinct that something was wrong, but I was already an outcast and they were not in any kind of mood to hear what I had to say.

Thankfully, Mike took my side and was able to persuade the command that it was worth investigating. We called on the radio but got an odd reply from someone on the boat that they couldn't see our ship despite us being within a mile or so of the sailboat. Once again, Mike took my side and pressed again for us to investigate. Reluctantly, the captain agreed to send over a boarding team. Another friend of mine led the boarding team and within minutes of him climbing aboard, he radioed back for the corpsman to come over. Shortly thereafter, with our corpsman transported over via the small boat, he relayed that the sole occupant of the sailboat was in diabetic shock and would need a medevac.

Sailing from Canada to Massachusetts, the owner had been seasick and unable to keep much food or water down for the past few days. He'd been sick down below and was in bad shape. As the sun was setting, an H-60 Jayhawk from the Coast Guard Air Station in Cape Cod arrived to take the sailor back towards Massachusetts and to a hospital. This left us with his boat, a 35-foot Southern Cross sloop. His intended port was Provincetown, Massachusetts, which was another day or so to the west. Sensing an opportunity to get off the cutter and into some fun, I volunteered to sail his boat back for him. Amazingly the captain agreed and my roommate, a classmate and football player from the academy, volunteered to go with me. So that night, we provisioned ourselves as best we could with drysuits and some food and were motored over to the sailboat with *Tahoma*'s small boat. From there, I trimmed the mainsail as best I could for a westerly course and fired up the engine to give us a little boost along the way. We also found a small CD player and a few CDs to keep us entertained throughout the night. My roommate promptly got seasick himself and laid down in the cockpit to try and rest. This left me with my own thoughts as we sailed west in chilly air under the North Atlantic stars.

It was the first opportunity I'd had to make any sort of meaningful contribution. Had it not been for Mike backing me up, the command would have likely blown off my suggestion that we investigate further, and the outcome would not have been good for that lone sailor. From my time sailing, I had an appreciation for the vulnerability sailors feel when they're out to sea and at the mercy of the wind. Most importantly, I'd learned beyond a shadow of a doubt that a sailboat making no headway on the open sea meant trouble.

Tahoma shadowed us throughout the night to make sure I didn't get into too much trouble, and the following day we sighted land. Meeting up with a small boat from the Provincetown Coast Guard station, they towed the boat the remaining way into the harbor. Finally back aboard *Tahoma*, there was no mention of the previous day or what we'd done to help that sailor. It was back to the usual grind of standing watches and deskwork. While I felt a tremendous amount of pride for what we'd done, it was so quickly swept under the rug that it increased the disdain I had for that ship. It was foolish to have expected perfect leaders in the fleet. Just as had sometimes been the case at the academy, I was once again learning not so much who I wanted to be, but rather who I didn't want to be.

After two years on *Tahoma*, I received orders to Coast Guard Recruiting Command in Arlington, Virginia. This was not the assignment I'd hoped for. In fact, it wasn't even on my list of 20 or 30 jobs up and down the east coast that I'd asked for. The silver lining was that I had stayed in touch with the coaches and director of the sailing program at the academy, and they'd asked if I was available to work again at the waterfront as a safety officer on *Gem* for the 2005 ocean racing season.

My command on *Tahoma* was probably excited to let me go in late April when we returned from a patrol down in the Caribbean. I unceremoniously left the cutter fleet for good and made my way back to New London, this time staying in Munro Hall, the transient quarters that afforded far more creature comforts than Chase Hall. I easily slid right back into the groove of prepping *Gem* for summer racing. I spent the next 60 days very happily working seven days a week in a pair of khaki shorts and a polo shirt, taking care to avoid the inevitable sunburn from days spent out on the water. We raced in and around Newport as a new crop of cadets I'd never met before learned the ropes and battled their way through cold nights, squall lines,

Onboard *GEM*, 2005, coaching the ocean racing team

and ever-present fatigue from long watches and sail changes in the middle of the night. It also gave me an opportunity to look back at what I'd taken away from the sailing program. Even though I'd been disappointed by my time on a cutter, it was now easier to recognize that my comfort out at sea was a direct result not from *Eagle* or Nautical Science classes, but rather from the days and nights spent on the deck of *Rampage*.

I left New London bound for Northern Virginia with a smile on my face. Reporting aboard Recruiting Command, I was met by mostly friendly faces and my boss, a Lieutenant Commander, seemed genuinely interested in getting to know me. While I was not excited to be sitting in a cubicle at the age of 24, the relaxed atmosphere was a welcome change. I was given far more responsibility than I ever had on *Tahoma*.

Specifically, in my first week, my boss told me we needed to trim something close to $300,000 from our annual budget. He then told me he was leaving on a trip and would call me in a day or two to discuss it. With that, he was gone, and I was left to figure out not only what our budget was, but also what we were spending a few million dollars on each year. Compared to *Tahoma*, this was a vast sum of money. As the communications division officer on the cutter, I'd been told I had a budget of $1,600. On top of that, any time my guys asked for things like new staplers or pens, I was relegated to being the middleman as I tried to convince the operations officer to approve their request that amounted to little more than a hundred or so dollars.

I immediately went to work and put my questionable math skills to good use. Thankfully our budget was easy enough to sort with some basic addition and subtraction, and there was no need for algorithms or those pesky letters that had somehow held numerical values during my three semesters of Calculus. As promised, my boss called me from his hotel and asked what I'd come up with. We discussed trimming enough from a few accounts and reducing some sponsorships over the span of about 15 minutes. At that point, he asked if I could have it completed by the following week, which I promised to do. He hung up and I went back to work again with our accounting folks.

That introduction to administrative duties was far from exhilarating life-saving missions on the high seas, but the trust that he'd given me in my first week was enough for me to dive into the recruiting job headfirst. It was a three-year assignment and unfortunately not the kind of job that would set a junior officer up very well for a career in the Coast Guard. I'd discussed with my new command early on that I was very interested in applying for flight school and, with their full support, I began the lengthy process. On my first two attempts, I wasn't picked up. But the third time was a charm, and I left Washington D.C. after about a year and a half, making my way down to Pensacola, Florida to begin the next chapter.

After reporting aboard to the Coast Guard liaison office, which was appropriately in the lower office space of the old Pensacola Lighthouse, I wandered around the Naval Aviation Museum one afternoon while waiting on my class assignment. My brief time at Headquarters was unplanned but had served to give me far more confidence than I got from two years on the boat. At Recruiting Command, I worked for good people, and in a short time they trusted me not with making big decisions—which were ultimately made far above my pay-grade—but I had an opportunity to influence those decisions. My voice mattered, and I was able to influence some significant decisions on behalf of the organization.

Selected for promotion to Lieutenant shortly before leaving Headquarters, I was confident that I had the skills and motivation to make it through the nearly two-year training pipeline. There was nothing else that I wanted to do in the Coast Guard. Truthfully, I had begun interviewing for jobs outside the Coast Guard prior to being picked up for flight training. Had I not been selected, I would have likely separated once my five year obligation was up.

Yet here I was, in the cradle of Naval Aviation, about to embark on yet another journey. The first step was a weeding out phase called IFS, or Introduction to Flight Screening, where we were sent out to local civilian flight schools to spend 25 hours flying a small single-engine Cessna or something similar. I ended up at Pensacola Regional Airport, flying a Piper Tomahawk. Googling the airplane my first night after classes, the internet informed me that its nickname was the *Trauma-hawk* and that the plane did not recover from spins well. This didn't bother me much, as I was young and certain that nothing bad could ever happen to me.

I spent just under two months at Pensacola Regional Airport, where we were hurried through ground school and then into the plane. After the first few flights, I was sent out on a solo and I flew the plane by myself, with my instructor watching from the ground. A successful solo was measured simply by the number of takeoffs and the number of landings being equal. We then did some very short cross-country flying and a few more flights at night. Overall, it was an enjoyable two months where our uniform was a pair of jeans and a polo shirt.

Almost immediately, I was classed up for Aviation Preflight Indoctrination, which was a six-week grind of academics and survival training. Here, I studied with the same intensity, if not more, than I had at the academy. Tests began almost immediately, and any failure could result in rolling back to the following class. A second failure likely meant one was removed from training altogether and sent packing. I did well, keeping myself slightly ahead of the bell-curve among a mixed class from the Navy, Marine Corps, Air Force, and Coast Guard. My mindset had changed, and I was no longer just trying to survive. If I put in the time, I would one day earn my wings.

The third phase was called 'Primary' and we upgraded planes to the T-34C Turbomentor. I flew nearly 100 hours over the span of just under eight months, starting with straightforward landing pattern practice before progressing through several phases that culminated with aerobatics in a formation. In our first month or two, we completed a solo flight before moving on to stages focused on flying in weather, performing aerobatics and navigation both over land and over water.

The weeks focused on aerobatics were the most enjoyable, and we had three solo flights where we took the plane out for a little over an hour each time, looping and banking around the towering cumulonimbus clouds that hung over the Gulf Coast.

I'd take off with a guilty grin on my face, working my way west towards Mobile Bay before finding some clouds to rip around. Once close enough, I'd roll the plane on its side and pull hard to skim the side of a cloud in a hard banking turn, level back out, then find another target. In between, I'd take a quick breath before looking up, pulling hard, and keeping the wings level to fly the plane through loop after loop. We had roughly an hour and a half for each solo aerobatic flight, and I was in a deep state of pure bliss each time I touched back down at Whiting Field. To this day, those were some of the most fun flights I've ever had.

In between flights, there was an enormous amount of studying and lectures we had to attend. In addition to the tests we took, each flight was graded. From the moment we sat down to brief the flight, to the moment we walked away from the plane, we were under the watchful eye of our instructors. Each of them came from different backgrounds and had flown various aircraft in the fleet prior to their tours as instructors. Some of them came across as jerks, and others were fantastic to be around. Each time we landed, they'd fill out a grade sheet that fed into a larger grading scheme, the exact equation being a guarded secret. At the end of each phase, if our overall grades were too low, we'd be rolled from training. It was stressful, but the hours of studying and practicing paid off each time I brought the plane in for a landing and walked away from another successful flight.

Naval Flight Training was not without risk. Early in Primary, I had sat through a lecture given to us by a Major in the Marine Corps about our flight publications. During the lecture, he asked us for suggestions on things we thought could be improved in our flight manual. The flight manual was the bible of the T-34C, describing everything from systems integration to standard and emergency procedures. None of us in the room felt comfortable enough with the content of the manual to say anything, but Major David Yaggy persisted and said we weren't leaving until he had a few suggestions from the group.

I took a breath and raised my hand. Pointing at me, Major Yaggy asked what I had to offer. I can't remember what it was that I brought up, but he wrote it down in his notepad, thanked me, and then produced a six pack of Guinness from behind his podium, walked it over to me, and dropped it in my lap. He then asked if anyone else had suggestions and the room erupted with a flurry of raised hands. The rest of the booze he gave out that day was comically sub-par compared to the six pack of Guinness, but he got what he needed from the class. Months later, and shortly after I left Training Squadron Six bound for Corpus Christi, where I'd enter the next phase of training, Major Yaggy and a student were killed in a crash.

At Training Squadron Thirty-One, I flew a modified version of the King Air 90, a svelte twin-engine turboprop airplane that could climb high, cruise fast, and was, in my humble opinion, much easier to fly than the T-34C. While I had enjoyed things like aerobatics and formation flying, in the King Air we didn't have to wear a helmet or survival vest, which made sitting in the cockpit far more enjoyable. I flew roughly 80 more hours in the King Air, joking at times that the second engine was pointless given that the instructors failed the second engine on almost every flight, until they were reasonably certain we had the skills to bring the plane back and land on one engine. For my time in Corpus Christi, I rented a place with a good friend of mine, and while still stressful at times, we enjoyed the six or seven months of training and did our best to account for the party deficit we'd incurred from our days at the academy. I winged in July of 2008, leaving with orders in hand for Air Station Clearwater

Florida, where I was beyond thrilled to have an opportunity to fly the C-130H Hercules.

Chapter 17

After several months of additional training, and now back with the Coast Guard, I was designated a co-pilot and cut loose to enter the duty rotation for the duration of my four-year assignment. Coast Guard Air Station Clearwater was a large unit with 60 or 70 pilots supporting the C-130 and H-60 flying. On any given week, we'd stand a few 24-hour shifts at the Air Station, with the expectation that we could be airborne in less than 30 minutes for a search-and-rescue case. In addition, both the C-130s and H-60s were continuously deploying throughout the Bahamas and Central America. Any Coast Guard Air Station is an exercise in controlled chaos, with planes coming and going at all hours of the day. It can be tough for a family given the constantly changing schedules, but I quickly fell in love with the lifestyle.

My experience from sailing again came into play during one of the more notable law enforcement cases on which I flew. In January of 2009, a troubled father had kidnapped his son and fled on a sailboat from somewhere in Tampa Bay. The mother had contacted the authorities who had quickly come to us for assistance. In many cases, law enforcement and search and rescue were intertwined missions.

The father had mentioned something to a friend about heading to New Orleans, so our initial search was in that direction. When flying over the Gulf of Mexico, the winter weather can be unpredictable and surprisingly cold. As we flew northwest from Clearwater towards New Orleans, our camera and radar operators were busy scanning the water for any signs of his boat. I thought back to those cold nights I'd spent aboard

Rampage and commented to the crew about the unfavorable winds from the northeast and the deteriorating weather we were seeing on our current track. He was far more likely to take the easier path and run with the wind which would have taken him west.

Anyone so desperate and delusional that they would kidnap their kid was far more likely to make a run for somewhere like Mexico with the false hopes that they could hide from the law. We talked as a crew and eventually doubled back to the south towards the western edge of our assigned search area. Not too long after that, our radar operator painted a small contact tracking southwest that was outside of our designated area. However, it matched our working theory that he would take the easier path and run with the wind. The small radar track showed a straight course towards Mexico, and as we diverted in that direction, we were confident that we had found them.

Once overhead and orbiting from several thousand feet above, our camera operator zoomed in, and the radar contact was a match for the description of the boat we'd been given. We brought the throttles back and orbited for several more hours, relaying back to our unit and the command center running the case. A second C-130 was launched that relieved us and orbited the sailboat all night as it continued slowly towards Mexico. The following morning, Clearwater launched an H-60 with a police negotiator onboard to rendezvous with the boat. At the same time, a cutter had gotten underway from Tampa and sped through the night to conduct a boarding to retrieve the child, Luke Finch, and return him to his mother, anxiously waiting in Tampa.

With a C-130 overhead, the H-60 arrived the following morning, coming into a hover as the small boat from the cutter pulled alongside and rescued Luke. With police onboard in addition to the Coast Guard crew, the father was detained, and all were transferred over to the cutter for transport back to Tampa. Luke was given free reign of the boat for the transit back home, which he enjoyed as the young boy quickly made friends with most of the crew. The cutter also took the father's sailboat in tow for the trip back to Tampa Bay. Overnight, and while being towed, the sailboat developed a leak and ultimately sank before the sun was up the following morning. Had we not found Luke and his dad, their odds of survival on that sinking boat would have been poor. I can't be sure that my suggestion about the path a cold and desperate sailor would take made the difference, but the discussion that followed solidified the idea for the entire crew and reaffirmed our decision to deviate west from our assigned search to investigate a lone radar contact. Any successful search-and-rescue mission is the sum of dozens of small decisions on the part of everyone involved. For Luke Finch, I like to think that my nights spent shivering on the deck of *Rampage* paid off in his favor.

The second case that left an impression on me was during Tropical Storm *Claudette* in the summer of 2009. We launched ahead of the storm to make radio broadcasts along the Gulf Coast about the overnight strengthening of the storm. It was a benign flight, but when we returned to Air Station Clearwater, we were quickly launched again for a distress call from a fishing boat, the *Big Jim II*, who had been caught off guard by the rapidly deteriorating weather. We took off again and flew an hour or so northwest of Tampa Bay, where we were able to locate the boat that was now caught in one of the more intense bands of the storm.

Establishing communications with the two men on the boat, we descended to roughly 500 feet. Every few seconds, and through the clouds, we could catch a glimpse of the surface, which was a tangled mess of churning whitewater and waves. It was almost impossible to spot the *Big Jim II*, but they were at least able to talk to us on the radio. We initially prepped to drop them a raft, as the crew had indicated that they might lose the boat to the seas. In the cockpit, we talked, and our navigator suggested we try to give them a bearing to steer to get out of the worst of the weather. It was a safer course of action than dropping them a raft, which would likely be blown away from them in a matter of seconds in the 60-knot winds.

Anytime we dropped survival gear to a boat, there was a risk of the survivors entering the water to retrieve it if we missed. Staying with the boat is almost always the safest thing a mariner in distress can do, and we didn't want to risk making a bad situation worse. I radioed the captain of the *Big Jim II* and asked if he was able to steer a westerly course. I can clearly recall his reply, panic coming through in his voice as he yelled over the wind that he feared the boat might take a wave broadside and be swamped if he turned. I pressed him again, recommending he turn west from his northerly course that he was holding, which was keeping his bow into the seas. At some point, he radioed back that he would try. That was quickly followed by another call that he'd taken a wave over the starboard rail and his deck was swamped. Over the wind, he yelled into the radio that they were both going to abandon ship. I keyed up my radio one more time and pleaded with him to hold his course. At the same time, the crew in the back of our plane went to work setting up to drop the raft, fearing that the captain would lose his boat at any moment.

Agonizing seconds passed before he radioed again that he was now holding a westerly course. We stayed overhead, unable to see the *Big Jim II* through the low clouds and rain squalls, but we had a good return on our radar and were able to keep constant communications going with him. Several hours passed and we orbited above, passing updates periodically back to the command center in Mobile that was running the case. Our navigator kept a big picture view of the storm, and we passed incremental course changes to the *Big Jim II*, until the captain radioed that he was able to navigate on his own and would be waiting out the storm until it made landfall. He would then make his way north again. With the storm quickly tracking north and the *Big Jim II* clear to the west, we headed home, our fuel tanks having little more than enough for the hour-long transit back to Clearwater.

The *Big Jim II* was not just a fishing boat; it was also someone's livelihood. Losing his boat would have been a significant setback for that captain's family. Losing the captain and his crew would have been even more catastrophic. In those conditions, there was a very real chance of both. He had asked us at some point when we first arrived on-scene if we were a helicopter, hoping that we'd be able to hoist him and his crew. The Command Center in Mobile relayed over the radio that they were unable to launch a helicopter due to the winds, and we were the only available asset to get on scene. In the middle of a tropical storm, despite the turbulence and noise in that moment, I smiled to myself for a second, knowing that I'd soon take full advantage of that radio transmission to taunt my helicopter brethren back in Clearwater.

But with jokes aside, the gravity of the situation was not lost on me. For the *Big Jim II*, it was a terrifying experience. I'd been in storms before, at times ignorant of the risks we faced racing a small sailboat on the ocean. But in many ways, I could better appreciate what the crew of the *Big Jim II* was experiencing. It made the entire case much more real to me. They weren't just voices on the radio, they were sailors in trouble and in need of help. There was no singular action on the part of any of us on that crew that made the difference in this case. Rather, it was the open dialog and quick-thinking that led us to abort the attempted aerial delivery of a raft and focus on giving the *Big Jim II* a course to steer. Our navigator had the best radar on the plane and the certainty in his voice led us to get behind his suggestion. We were confident in our actions and perhaps the captain could hear that confidence over the wind, and that was enough to keep him in his boat for a few more minutes. Those seemingly trivial details had made the difference.

My first tour was not without tragedy. I'd gone through flight school with several classmates from the academy, and most of us had made it a point to get together socially on the weekends. Adam Bryant was several weeks ahead of me in each phase of training, and I would often chat with him at parties as he was a good source of what I could expect in the coming weeks. He finished Primary a few weeks before I did and then went through Advanced training a class ahead of me. I had one Navy instructor in Advanced who would often tell me I'd flown a good flight, but that it had not been "Adam Bryant good." Adam was the type who would come to any party anyone hosted, but also the type of guy who would say his goodbyes well before midnight, most often with some comment about how he had a lot of studying to do the next day. I lacked the same discipline with which Adam approached studying on his free time. After flight school, he completed his C-130H transition course just prior to mine, before reporting to Air Station Sacramento, California. In October of 2009, a C-130H Hercules, tail number 1705, launched on a search-and-rescue case for a missing boater near the Catalina Islands off southern California. During their nighttime search, the aircraft collided with a Marine Corps helicopter that was conducting an unrelated training mission.

Both crews were killed. My roommate, both from the academy and our time in Corpus Christi, called to tell me Adam had been the copilot.

I left Air Station Clearwater in the summer of 2012, bound for Air Station Elizabeth City, North Carolina, where I transitioned to a newer version of the C-130, the J model 'Super Hercules.' Our area of responsibility covered most of the Atlantic Ocean, to include the Outer Banks of North Carolina, often referred to as the 'Graveyard of The Atlantic' due to the numerous shipwrecks that rested on the bottom. Dating back centuries, the Outer Banks were known as a treacherous stretch of coast where the Gulf Stream ran headfirst into the shallow banks off Cape Hatteras. Even with the end of the age of sail, ships were often caught off-guard by the fierce seas that could develop in a short time due to the hurricanes and nor'easters that frequented that stretch of ocean.

In the fall of 2013, I launched one evening to search for a sailboat that was overdue off the coast of Virginia. A gentleman had intended to purchase older sailboats in and around the New York area and then sail them down to Florida to turn a profit. His girlfriend had reported him overdue for a check-in and that quickly made its way to the command center in Virginia. It was late in the evening when we took off with an assigned search area that was based upon his intended track towards Florida. There were high winds from the northeast and the sea state was deteriorating as we made our way east over the Atlantic and then north from Elizabeth City. Adding to the misery of being in distress on the water, it was an unseasonably cold night for early fall.

We were told that he was the sole occupant of one boat and was towing a second behind him. I asked the command center if it would be all right if I deviated from the assigned search to check the shoreline of the eastern shore of Virginia, thinking that if he was in distress, he would head for land rather than tough it out in the deteriorating conditions at sea. The command center was indifferent, but ultimately agreed that I could take our first leg up and down the eastern shore before proceeding further out to sea. With night-vision goggles, we flew low over the uninhabited islands of the eastern shore, scanning for any signs of the boats or a person. Just inside a shallow bay near Hog Island, we sighted two masts, tucked just inside the north edge of the island.

It was just after midnight when we made several passes at about 200 feet to get his attention. Unsure of his condition, we then dropped a can full of survival equipment via a parachute to the beach next to his boats. At that point, we were able to get ahold of him on the radio and he claimed to not be in distress—despite having two boats nearly aground on an uninhabited island after midnight during a particularly stormy night. We coordinated for a small boat from a nearby station to proceed to his location and assist in whatever way they could. We then relayed this to the sailor on the beach. While the particulars of his situation weren't great, he was not in any immediate danger and the small boat would be able to get him back to civilization. Thinking our job was done, we climbed back up and headed home.

The following morning, we found out that he'd headed back out to sea prior to the small boat reaching him. He'd likely guessed that the Coast Guard would not let him continue on his voyage to Florida. With money at stake, he'd slipped out under the cover of darkness and headed south once again, determined to find his fortune. Days later, he was once again on the Coast Guard's radar when he ran aground just south of Oregon Inlet on the Outer Banks. Having once again encountered bad weather, he'd been pushed ashore while apparently catching some sleep. Both his boats ran up on the beach and he tried in vain for several weeks to save them, but to no avail.

In true Outer Banks fashion, the locals tried to help him, but his boats were damaged beyond repair. One was quickly destroyed while the other, *Belle*, sat for several years, slowly eaten away by the tides, winds, and shifting sands of the Outer Banks until she finally disappeared for good. Several years after *Belle* had run aground and been left for scrap, I parked at Oregon Inlet and hiked a few miles down the beach with my four-year-old daughter on my back. I told her that we were looking for a pirate ship which kept her interested in the afternoon adventure, and once we found *Belle*, I took some time to reflect on both the recklessness and the bravado of the long-gone sailor who had tried unsuccessfully to bend the will of the ocean.

After my assignment in Elizabeth City, I transferred to the Aviation Training Center in Mobile Alabama, where I took over the day-to-day operations of the C-130J Training Division, or STAN team. Our primary job was to conduct the initial co-pilot training for pilots reporting to the C-130J. What made the job interesting was that ATC didn't have any C-130s in Alabama. Rather, we were co-located in Elizabeth City and had an agreement where we used the Air Station's aircraft to conduct our training flights. ATC also lacked a simulator for C-130Js, so we spent a considerable amount of time travelling across the country using simulators at a handful of Air Force and Marine Corps bases. This meant that my boss was almost 1,000 miles away from me, which was good and bad, depending on the day.

In addition to training new pilots, we were the default source of crews for testing new equipment on the plane. When I took over, we quickly became involved in conducting test flights for a new mission system. This consisted of a belly-mounted radar on the plane, a camera mounted on the nose, and a host of long-range communications and data sharing capabilities. I had several mission-system operators working for me at the STAN team and together, we became heavily involved in the initial operational testing. One C-130J at the Air Station was upgraded to the new system, dubbed *Minotaur*, and we had *carte blanche* with that plane.

In August of 2017, I took off on what was supposed to be a routine training flight. I had one mission system operator with me and one of my pilots who had not flown for over a month, which triggered a warm-up flight requirement for him to re-establish his currency in the plane. In the back, I had one more crew member to keep an eye on things in the cargo compartment. For search-and-rescue missions, the required crew would have been a minimum of seven, but for a training flight, we would make do with as little as four. Shortly after takeoff, we got a call from a nearby command center asking if we were able to divert and look for a missing sailor in the Pamlico Sound, off Ocracoke Island. A squall line had blown through unexpectedly and left the sailor out of sight of his friends.

This case would have normally required a full crew and mission-capable plane. I had with me a partial crew, a prototype mission system, and a non-current pilot sitting next to me. Nevertheless, I didn't hesitate to turn the plane around and make our best speed towards the Outer Banks. Enroute, I asked my sole mission system operator if he would be comfortable powering up the prototype system. He replied that he was already working on it. I then polled my crew, asking if anyone had objections to diverting for SAR even though we were not fully within the rules. In true Coast Guard fashion, we all wanted to get it done.

We arrived on-scene just as the storm was moving east over Ocracoke. Dropping down to 1,000 feet, we scanned as best we could and flew an initial search in the area. Spotting a small sailboat from a C-130 was tough, and we were almost entirely dependent on the mission system radar and camera to find him. In a short time, my mission system operator vectored us in towards a small and intermittent contact on his radar. Normally a second operator would work the camera, but he was on his own to work both. Despite that challenge, he had the camera zoomed in moments later on a small sailboat drifting several miles out into the Sound. Laying prone on the boat we could see a single sailor, and as we passed overhead, he waived one arm up at us, but didn't move much more than that. His lethargy did not give us a good feeling.

Once we called back to the Air Station, they promptly launched an H-60 Jayhawk. It arrived on-scene within 30 minutes, and we vectored them in for a recovery. The rescue swimmer was quickly lowered down via the hoist cable and from there he unhooked to free fall the last few feet down into the water. We watched as he then awkwardly stood up in about four feet of water. Dressed in a wetsuit with long swim fins, he struggled to shuffle over to the waiting survivor. We laughed a bit, realizing that the distressed sailor could have, in theory, literally stood up and walked most of the way back to land. The reality was far different from his vantage point. At sea level, he had long ago lost sight of land, having been caught in a line of weather and no doubt disoriented by the wind and rain. For all he knew, he was drifting further out to sea. His choice to stay on the boat was a smart one also in that had he abandoned it and tried to swim or walk to shore, it would have been even more difficult to find him.

Hoisting the sailor, the Jayhawk radioed to us that the he was having chest pains and difficulty with his vision. The plan quickly changed from bringing him back to Elizabeth City to getting him to the closest hospital on the Outer Banks. We stayed with the helo for some time, orbiting overhead and re-laying communications for them about the condition of the pa-tient while also working clearances through some areas used by the military for live-fire exercises. Thankfully, the survivor made a full recovery.

After landing, I called my boss and laid out the full list of rules that I'd broken on the flight. I had an unqualified pilot with me, an incomplete crew, and a still-uncertified system that we'd powered up and employed outside of the test protocol. The flip side was that we had also saved someone's life. My boss laughed into the phone and told me to go home and get some rest.

Chapter 18

The motto of the Coast Guard Academy is *Scientae Cedit Mare*, which affirms to its graduates that "the sea yields to knowledge." After a career on and over the ocean, I would argue that this is a flawed statement. The caveat would be that there is no guarantee that the sea will ever yield to anything. From days sailing as a cadet, to the years I spent as a pilot flying low over the world's oceans, I have seen the sea both at its best and worst. The ocean is a thing to behold, to admire, and most importantly, to respect. The ocean can take and the ocean can give many things. With time, anyone other than a fool comes to appreciate the power she holds.

The mission of the Coast Guard Academy is rather long but begins by pledging that its graduates will leave with a "liking for the sea and its lore." This is, in my humble opinion, a far more appropriate goal than to graduate with an expectation that the academy will teach us the skills to tame oceans. Confidence is an attribute, but arrogance will quickly get you killed. Ensuring that its graduates develop both confidence and humility in the face of adversity is a lofty goal, but also a noble one. I left the academy unsure of how the process was designed to work and equally unsure of whether it had worked for me. I had most certainly been broken down, but far less certain of what I'd been built back up to.

There was no formal curriculum that guided the sailing program, and that is why I gravitated towards it with such energy. Given that blank slate, I felt as if I had a chance. Joining the Coast Guard was an opportunity to challenge myself in an environment that I loved. The Coast Guard's maritime roots were drilled into us beginning on that first day in July of 1999, but no lesson was as immediate and real as those that I learned on and in the water. From my first days on dinghies sailing around the Thames to summers spent on the open ocean, it was a seemingly random confluence of events that taught me the many challenges of simply being at sea. Quick thinking must at times be tempered by patience, perseverance is worthless if not accompanied by a keen eye that recognizes overwhelming odds and the need to change course, and risks are things to be carefully calculated against both the best and worst possible outcomes. And above all, when certain of the worth of your endeavor, never ever give up.

Karin Marley Simons and a friend tried unsuccessfully to smuggle cocaine into Halifax, Nova Scotia in late August of 2021. They arrived on a sailboat, but the authorities had been tipped off and quickly apprehended them both. Prior to their arrest, they'd burned the boat in an attempt to destroy evidence, but their efforts had failed. Karin had been injured by the ensuing fire. After his arrest, the authorities took him to a local hospital for treatment. This is quite a story in itself, but Karin's adventure was only beginning.

At some point, the 32-year old from Antigua escaped, barefoot and wearing only a pair of shorts and a hoody. What happened after that is not entirely known, but it's suspected that he commandeered the *White Wing,* a 24-foot sailboat from a local marina. Surveillance footage confirmed that the boat was motored out of the basin in the dead of night. Days later, it was found washed up on the rocks off Stanbro, to the east of Halifax. The boat was too damaged to be salvaged, but according to the owner, who looked at photos of the boat, Karin "left the night of Hurricane Ida. It was blowing out of the southeast like forty-five knots. So, it was a strong wind... My gut feeling is that he rode the boat right to the bitter end and then got off on the rocks. He went for a hell of a ride to survive that."

Canadian authorities could not be certain that it was Karin who had stolen the boat, but they assumed it was, given his certifications and experience as a deckhand and sailor from Antigua. According to news articles, it was assumed that Karin had perished when the boat ran aground. Yet, days later, another boat, *Secret Plans,* was missing from a marina in Halifax. Her owners only found out when they were notified that their emergency beacon was transmitting hundreds of miles south and in the center of yet another storm, Hurricane Larry, which was making its way to the northeast by way of the southern coast of Halifax. *Secret Plans,* as it turned out, was a more capable boat and provisioned enough that one might have had a chance of making the days-long journey back to the Caribbean, had it not been for a hurricane standing in the way.

With *Secret Plans'* emergency beacon activated, the Canadian government sent two aircraft to reconnoiter the area, but both had turned around in the face of hurricane-force winds and seas. The Canadians did what every country does when the weather is god-awful and exceeds their own capabilities: they called the U.S. Coast Guard. And, as is always the case, we agreed to give it a try. Back in Elizabeth City, I was nearing the end of a 24-hour duty period when the alarm went off. Hustling to our command center, I got the basic details that a boat had turned on its emergency beacon in the middle of a hurricane off the coast of Halifax. At that point, I didn't know that the sole occupant of the boat was running from the law. It didn't really matter, however, as asking about the particulars of a sailor in distress wasn't something we did. Someone was in trouble, and we were the lucky ones to go try and help.

With my crew, we hustled to the aircraft, but I remembered reading something about that plane's weather radar not working so well. I reluctantly went back to the hangar and asked for a second plane with a working weather radar. I was reluctant because this would delay our takeoff, but I also felt strongly that if I was going to fly into a hurricane, I wanted to see the radar returns so I could pick my way around the worst of the storm. Without a weather radar, it would have been like a blind man stepping into a boxing ring to square off with a heavyweight.

Soon, and with a different plane, we took off and pointed northeast for what would be a three-hour flight before reaching the outer limits of Hurricane Larry. Once there, we descended into the first of the outer bands and picked our way around the worst radar returns. An initial search in the vicinity of a last known position of the emergency beacon turned up empty. To search, we normally put the aircraft on its autopilot so we could focus most of our energy on looking for whoever was in distress. However, in the 100-mile-an-hour winds and turbulence, the autopilot couldn't control the plane. Making matters worse, we hit some kind of downdraft that instantly knocked the plane about 200 feet below our initial altitude of 1,000 feet. It took me some time to regain control of the aircraft and in those few seconds that passed, we were blown miles away from the assigned search area.

I had two thoughts. First, we were incapable of conducting any kind of meaningful search. All my efforts were going towards keeping the plane out of the worst of the weather and, more importantly, out of the water. Second, if we were to find the boat, our procedures would be to drop a life raft to the survivors from an altitude of 200 feet. I'd just lost 200 feet of altitude in a second or two, which meant that if we were much lower and encountered another downdraft, our own chances of survival quickly diminished.

I recovered and climbed to regain some altitude. And at that point, I looked out the side windows of my plane and down at the seas. It was late afternoon, and the sun would be down in another hour. But the ocean below me looked like nothing I'd ever seen before. I'd flown over heavy seas and seen big waves breaking over themselves, with whitewater tumbling then disappearing back into the blue. But in this storm, there was almost no blue left in the ocean. Rather, it was blurred shades of white and grey, the breaking waves being rolled over by the tops of other waves and the whitewater buried under more whitewater from even more 50-foot waves that were cresting in every direction. I took a moment to take it in. It was both a savage and strangely beautiful thing.

It was also not survivable. Climbing up again to get us even more distance from the ocean below, I told my crew we were headed home. I pointed the plane back to the southwest, towards North Carolina, some 900 miles away. There was nothing I could do. Less than a minute later, we got another call that the emergency beacon was still transmitting, now from a different location that was outside the worst bands of Hurricane Larry. With some reluctance, I agreed to go look. We transited some distance to a new location and descended back down to one thousand feet, still concerned that we might hit another downdraft. Settled on a track that took us south, we found that the winds in this new area had subsided to around 75 knots. The seas were still near 50 feet and as I scanned my surroundings, I thought once again that these conditions were not survivable. Whoever had activated that beacon was likely gone and the beacon was left floating in a sea of debris from what once had been *Secret Plans*.

Then, my copilot called out over the intercom and pointed at the same time towards the horizon in front us. I looked forward as two red flares popped up and were quickly caught by roaring winds before being blown back down to the surface, where they both extinguished. I whispered, "Oh. My. God." It was unmistakable. Karin was still alive, and he was in the water not more than a mile or two in front of us.

The ocean is always moving, sometimes just a little and sometimes a lot. This makes it incredibly difficult to see something or someone and then maneuver a plane to fly over it. With nothing to reference other than a gray horizon and a sea of whitewater, it's all but impossible to find anything or anyone, even with the C-130's very capable radar and infrared camera system. Making matters worse was the incessant salt spray that pelted the windows of our plane. That salt was crusting up over the glass, and it was becoming increasingly difficult to even see outside the airplane.

Still, there was someone in the water underneath us. We circled the area for some time, unable to relocate the source of the two flares. Nightfall was fast approaching and would make searching that much more difficult. However, we had night vision goggles with us and as darkness set in, if Karin had a flashlight, a strobe, or any kind of bright light with him, we'd be able to see him with our goggles. Twenty minutes or so passed and we continued circling the area before it was dark enough to put our goggles on. From the cockpit, we scanned all around, certain that an experienced sailor like Karin would have some kind of light with him. Clearly he'd seen us from the water and knew we were there to find him. With my jaw clenched, I told myself it was only a matter of time before he would signal to us, and we'd have a better idea of where he was.

Minutes turned into an hour with nothing. We were close enough that we could intermittently pick up the signal from his emergency beacon. It would transmit from the crest of a wave, and our mission system would beep to tell us that we were close. But with each passing wave under us, the beacon would drop back down into a void and disappear. We tried time and again to triangulate the exact position, but it was to no avail.

As a crew, we gave it our best guess and dropped a flare from the plane down to the water. It would burn for the better part of 45 minutes and perhaps at least give us some reference for what the direction of drift of the ocean was. Relocating that flare took us almost ten minutes. The trough of those 50-foot waves had swallowed the flare, and once we finally did relocate it as it was pitched over the crest of another wave, we were miles from the initial position where we'd sighted the flares from Karin. The ocean was moving in ways I'd never seen.

With someone in the water, we had opened the ramp of the plane in the back and two of my crewmembers were strapped in and searching for any signs of life. My mission system operators were also straining with both the radar and the Forward-Looking Infrared Camera to find any signs of life. I took it all in, the efforts of my crew, and the risks we were taking just to fly in those conditions. Several hours had passed and we still had to get the plane almost 900 miles back home.

I hated it but had to admit that continuing the search was hopeless. Karin was an experienced sailor. His two flares had signaled that he'd seen us. He knew we were there, and if he had any way to mark his position, he certainly would have. Our aerial delivery procedures accounted for winds up to about 70 knots. Even in the improved conditions, we were outside our parameters. Worse still, I knew that if Karin was still alive, at that point he had drifted miles from where we'd first seen him. In truth, we had no idea where he was.

Looking up, I scanned our fuel gauges, something I'd neglected to do for some time. If we stayed much longer, we ran the risk of not making it home and having to divert somewhere else. There would be another search in the morning, and the plane I had was the only available aircraft for that follow-on search. If we didn't make it home, I would have no doubt landed somewhere else, but the only chance Karin had at this point was to survive the night and be found with the benefit of daylight and calmer seas in the morning. I reluctantly told my crew we were headed home.

We closed the ramp and door in the back of the plane and climbed away. I was relieved to get some distance between myself and the hurricane, but as we ran checklists and calculated the fuel we needed to get home, I thought that Karin was still watching us from somewhere in the dark ocean below. Our plane was lit up, every light that we had was turned on to try and get him to signal to us, but it had been to no avail. If he was still alive at that point, he was certainly smart enough to know that he wouldn't be rescued that night.

That he was even able to shoot two flares into the sky when we'd first arrived was testament enough that the young man from Antigua was a sailor through and through. That he'd been smuggling drugs was not to be commended, but the tenacity with which he'd formulated his escape plan told me that he was a fighter. Despite hurricane force winds and 50-foot seas, Karin had not given up. He'd fought off the ocean's overwhelming strength and stayed alive long enough for us to find him. Yet, even with all of that, there was little we could do to save him.

We landed back in Elizabeth City near midnight and shut the plane down. I reminded my crew that two Canadian planes had turned around before even reaching the search area and that, despite our failure to save Karin, we'd all taken considerable risks in giving him even the smallest chance of rescue. I drove home exhausted, both mentally and physically. Thinking back to my first tour flying, I remembered the words of an instructor I'd looked up to. He had told me that the C-130 can take more of a beating than a human can. I contrasted that with the plane's autopilot kicking off and the downdrafts that had hurled us towards the ocean. It occurred to me that the plane had tried in her own way to tell us that enough was enough. And yet my crew had persisted. No one on that plane had asked to go home. No one had questioned, not even once, why we were there, at night, in the middle of a goddamned hurricane, trying as best we could to save the life of a complete stranger.

We failed to save Karin Marley Simons. There would be no recognition for our efforts, no medals on our chests, but on that mission, we'd given it our absolute all. I reminded my crew, all of them younger than me, of that fact because I worried about the emotional toll it might take on them in the years to come. I had found that in my later years of flying, it was much more difficult to brush such things aside. We left a sailor to die, but no other crew in the world took the risks we did to try in vain for some better outcome. I drove home incredibly proud of that effort.

In the days that followed, news articles popped up about Karin Marley Simons and the unbelievable story of his escape from the Canadian authorities. Another plane launched early in the morning and flew over the same ocean that we'd scoured the night before. There were no signs of Karin or *Secret Plans*. It took me a day or so to recover, both physically and mentally, before I felt like I was back at full strength. From my house near Chic's Beach, the same beach where I'd spent my last night as a civilian before making the trip north to the Coast Guard Academy, I took my dog for a long two-hour walk. In the lee of Cape Henry to the east, small waves lapped against the shoreline as my German Shepherd ran back and forth chasing seagulls from the sand and into water. I watched those small waves dance up to the shore, break over themselves, and roll gently back into the bay. Nearly 1,000 miles from Karin and whatever remained of *Secret Plans,* I felt somehow connected to him. At my feet, the Chesapeake Bay ran east out to the Atlantic and then north, 1,000 more miles to the southern coast of Halifax. Somewhere out just past the horizon, I was certain Karin was still there, holding a southerly course for home.

A liking for the sea and its lore.

About the Author

Brian Boland is a 2003 graduate of the United States Coast Guard Academy. After an assignment at sea and a brief tour in Arlington, Virginia, he completed Naval Flight training in 2008 and spent the rest of his career accruing  over 5,000 hours in the cockpit of Coast Guard C-130s. Brian holds a Master of Arts in Military History from Norwich University and somehow managed to retire from Active Duty in 2024 with the rank of Commander. During his free time, he can most often be found on or in the water around Virginia Beach hanging out with his awesome daughter, Elli.

Milton Keynes UK
Ingram Content Group UK Ltd.
UKHW020009010824
446405UK00010B/130/J